Organizational Facets of GIS

Organizational Facets of GIS

Edited by
Stéphane Roche
Claude Caron

ISTE

WILEY

First published in France in 2004 by Hermes Science/Lavoisier entitled *Aspects organisationnels des SIG*

First published in Great Britain and the United States in 2009 by ISTE Ltd and John Wiley & Sons, Inc.

ISTE Ltd
27-37 St George's Road
London SW19 4EU
UK

www.iste.co.uk

John Wiley & Sons, Inc.
111 River Street
Hoboken, NJ 07030
USA

www.wiley.com

Library of Congress Cataloging-in-Publication Data

Aspects organisationnels des SIG. English
Organizational facets of GIS / edited by Stéphane Roche, Claude Caron.
p. cm.
Includes bibliographical references and index.
ISBN 978-1-84821-085-1
1. Geographic information systems, 2. Land use--Planning. 3. Regional planning. I. Roche, Stéphane. II. Caron, Claude. III. Title.
G70.212.A75 2009
910.285--dc22

2009016651

British Library Cataloguing-in-Publication Data
A CIP record for this book is available from the British Library
ISBN: 978-1-84821-085-1

Printed and bound in Great Britain by CPI Antony Rowe, Chippenham and Eastbourne.

Table of Contents

Foreword

In the decade and a half since the release of the first popular Web browser (Mosaic, released by the National Center for Supercomputer Applications in 1993), the Internet and electronic communication have had a profound and still accelerating impact on society, at least on the well-resourced side of the digital divide. It is salutory today to read some of the literature of the early 1990s, and to note its almost complete failure to anticipate these impacts. In the introduction to the second edition of *Geographical Information Systems*, published in 1999, my co-authors and I noted our lack of even a single reference to the Internet in the two volumes, 56 chapters, and almost 1,100 pages of the first edition (Maguire, Goodchild and Rhind, 1991).

The first Web maps appeared within months of the release, allowing users armed with nothing more than a copy of Mosaic to display and interact with a map instantaneously by contacting the Xerox Palo Alto Research Center's server. The use of the Internet to provide easy access to geographic information (GI) grew rapidly, as exemplified by early repository projects such as the Alexandria Digital Library (www.alexandria.ucsb.edu) and later by a flood of more sophisticated geoportals (e.g. the Geospatial One-Stop, www.geodata.gov). Whereas early teaching in geographic information systems (GIS) had required students to go through the gruelling and mind-numbing exercise of hand digitizing, today that content has been almost entirely dropped from the curriculum, and students are now trained in how to find their data on the Web.

But like the "horseless carriage" of the early 20th century, this world of Web-based GI search and retrieval merely improved on previous arrangements, preserving almost all of their institutions and metaphors. By the turn of the century one could be forgiven for thinking that the Internet and Web had simply allowed us to do the same things better, sometimes more cheaply and quickly, but almost always with less skill and effort. The institutions that had evolved over decades remained in place, while the new technology placed more emphasis on the need to share GI among them, a recurring theme in the chapters of this book.

A far more fundamental change began to impact the world of GI in the appearance of wayfinding sites, initially popularized by MapQuest and later followed by a host of imitators. While these sites also provided access to GI, they functioned not as libraries ready to distribute data, but as services that processed their data to address queries posed by users. The functions of address matching (finding the geographic coordinates of a street address) and optimum routing (finding the best route from an origin to a destination) had long been available to professionals trained in the complex user interface of a GIS, but MapQuest made them available to all, and via a remote and well-maintained server rather than from the user's own desktop or laptop machine. More and more such granular GIS services became available online, providing cheaper, more reliable, and more accurate answers than most users could afford to provide for themselves. Eventually the approach became known as *service-oriented architecture*, and the use of such services as *cloud computing*. It also led to a profound organizational change, as users came to rely on companies for basic GI services, instead of the traditional combination of downloaded government data and local processing – and companies in turn found business models, often based on advertising, that could provide the necessary income stream.

One of the great triumphs of the Web has been its ability to provide easy access to a host of sources using very simple standards and specifications. HTML and a small number of graphics standards provided the initial impetus, replacing a Babel if not by a universal language or Esperanto (Chapter 2), then at least by selected specifications and plug-ins. In the GI world, online services now make the task of conversion between alternative methods of georeferencing (latitude/longitude, street address, placename, coordinate systems, etc.) trivial, so that information from multiple sources can be combined using location as a common key. The result has been a rapid proliferation of *mashups*, Web services that combine two or more sources to present information that would not have been available from either. For example, the Fundrace service (fundrace.huffingtonpost.com) makes it possible to create a local map of donors to political campaigns, by combining information from the national database of donors with a mapping service. By allowing Web developers to invoke its services from within their own, Google has established Google Maps as the basis for thousands of such mashups.

Another change of profound significance to GI occurred in 2005, driven in large part by the increasing sophistication of the graphics capabilities of standard computing hardware, which were driven in turn by the lucrative video-game market. Whereas it cost several hundred thousand dollars in the early 1990s to acquire a computer capable of dynamic three-dimensional display, by 2005 the average $1,000 desktop could perform almost as well. In that year Google released its Google Earth client, and for the first time the average user was able to display and manipulate the globe as a spherical entity. The centuries of Earth-flattening distortion that had been driven first by the necessity of paper communication, and later by the limited visualization capabilities of early computers, were suddenly over. Once more the effect was to reinforce corporations, and lessen the influence of the traditional public-sector GI institutions.

But the most profound organizational transformation was yet to come. By the late 1990s it had become commonplace for Web sites to solicit input from users, in the form of reviews (e.g. amazon.com) or advertisements (e.g. Craigslist), and to make this information available to others. Within a few years this type of practice had mushroomed to the *social media* we see today, the world of blogs, wikis, and other forms of user-generated content (UGC) that are often termed *Web 2.0* to distinguish them from the earlier domination of the Web by authoritative, top-down information flow. It was inevitable that this trend would impact the world of GI, because many forms of geographic information are intuitive, needing no particular expertise to collect, and because citizens are far more densely distributed over the Earth's surface than the experts who had previously dominated the production of GI. Satellite-based remote sensing and aerial photography are massively efficient techniques for GI acquisition, but there are many characteristics of places that such systems cannot "see". Who better than the average citizen to contribute, correct, or update such *volunteered geographic information* (VGI; Goodchild, 2007)? The term *neogeographer* (Turner, 2006) has been coined to describe citizens in this role, since they perform many of the traditional mapping functions of the expert geographer. Recently I have argued that the role of the neogeographer may not be limited to the more simplistic types of GI such as placenames and streets (Goodchild, 2009), but may complement or even replace traditional methods for creating all types or themes of GI.

Traditional GI institutions are often national (e.g. the national mapping agencies), oriented to authoritative production of data at multiple scales, and covering multiple themes. This new world of VGI is often focused on a single theme, and often local in its geographic coverage. For example, the 2007–2009 series of wildfires in Santa Barbara County stimulated the creation of several local Web-based services dedicated to rapid dissemination of GI during emergencies. Although this GI is generally not authoritative, it is often valued more by impacted citizens because it appears more rapidly than official data. Sites can be found dedicated to personal narratives about locations in Toronto (MurmurToronto.ca), variation over the US of the terms used to refer to carbonated beverages (popvssoda.com), or the creation of a volunteered global streetmap (OpenStreetMap.org). The institutions that sponsor or emerge in support of such efforts are themselves local, regional or global, and general or thematically based, depending on the context.

In many cases there is no existing institution that can fit the need created by such services. In Nova Scotia's rural Annapolis Valley, for example, the community like many others is actively engaged in numerous projects to create VGI, ranging from the mapping and transcription of gravestones to the recording of community genealogy. While such activities can flourish without any institutional framework, the need for continuity of the project and long-term preservation of its products will inevitably force one to be identified or created. Local governments typically lack the resources, regional governments are organized at too coarse a geographic scale, and while national governments may provide initial funding, that is unlikely to be sustainable in the long term.

In short, the profound changes that neogeography, VGI and Web 2.0 have wrought in the past few years, and the broader implications of social media, suggest a fundamental reorganization of society, a re-institutionalization, in which the older authoritative institutions either fade into irrelevance, or learn to collaborate with an empowered citizenry. This is the backdrop against which this book is being published, and it suggests that in ten, perhaps as little as five years the landscape of GI organizations that we currently see will have altered in dramatic ways.

Michael F. Goodchild[1]

Bibliography

M.F. GOODCHILD, "Citizens as sensors: the world of volunteered geography", *GeoJournal* 69(4): 211-221, 2007.

---

1 Center for Spatial Studies, and Department of Geography, University of California, Santa Barbara, CA 93106-4060. Phone +1 805 893 8049, FAX +1 805 893 3146, Email good@geog.ucsb.edu.

M.F. GOODCHILD, "Neogeography and the nature of geographic expertise", *Journal of Location Based Services*, 2009.

D.J. MAGUIRE, M.F. GOODCHILD, and D.W. RHIND (eds.), *Geographical Information Systems: Principles and Applications*. Harlow, UK: Longman Scientific and Technical, 1991.

A. TURNER, *Introduction to Neogeography*. Sebastopol, CA: O'Reilly, 2006.

Introduction

Western post-industrial companies are in the process of major changes. This shift is accelerating toward the not always understood concept of the society of information. On an individual level, the omnipresence of the Internet constitutes one of its main realizations. As for public organizations, territorial communities in particular, the concept of e-government is growing. It is based on the increased promotion of the circulation of public information to facilitate citizen access and increase the efficiency of public services. Within private organizations such as companies, information is a more valuable resource today than it ever was. It represents the foundation of a newly wealthy generation. The challenge for companies consists of following the accelerated rhythm of information flow, accelerated nowadays due to the effects of globalization, and to manage this intelligible knowledge.

The whole of society is facing radical transformations. The huge quantity of information that is produced and exchanged, combined with increasingly ubiquitous modern mass communication technologies (cell phones, personal digital assistants, etc.), generates a situation in which information resources constitute a motor for major force for social change.

Written by Stéphane ROCHE and Claude CARON.

Given the relatively recent character of these informational and technological changes, the societal context is still in a phase of intense movement. It is still quite difficult to provide a revealing description of these changes. In this context, the social and human sciences have a fundamental role to play in understanding the complexity and the impact of the communication and information sciences. Only a "systemic" grasp of human, organizational and social transformations caused by the information age can help us manage the risks associated with the often unrestrained and ill-considered diffusion of technology.

Information is the key component of these transformations. Even though this multi-faceted concept has several very different theoretical explanations, it concerns a large variety of fields. We can define it more specifically as a geographical component that is still being established today, so much so that we no longer speak only of information technologies (IT), but of geographic information technologies (GIT), which constitute a specific sector in IT. This sector is identified by different terms depending on the cultural or disciplinary context: geomatic, geographic information sciences (GIS). The field of GITs has experienced remarkable growth during recent decades, in particular with the arrival of microprocessing, the Internet, and the digital age.

The field of geographic information (GI) is characterized by specific dynamic practices: a large number of practitioners with varying professional backgrounds (land surveyors, topographers, cartographers, geographers, computer specialists, Internet users, etc.); a diversity of companies of different sizes and natures (companies specialized in data acquisition, geomarketing, GIS, image processing, etc.); numerous public organizations (government agencies for natural resources, cadastre, forestry, mining, land management, etc.); as well as a variety of users that increase with the diversification and applications of GI and the democratization of Geoweb and location technologies. This sector, intersecting by nature, also establishes its identity through a very active multi-disciplinary scientific community (universities, academies, institutes, and research centers in various fields including IT, geography, environmental science, etc.). The GI sector is a very busy community. We will assume here that the GI community constitutes the base of one of the last stages of evolution in the information society, what could be called the geographic information society (GI-soc).

However, the information society is not only a community of common interests and practices with systematic access to information as a common denominator. In more fundamental terms, it corresponds to the in-depth reconstruction of the operating methods of post-industrial Western companies as a whole, as centered on information use. Information in this society becomes the preferred engine of social relations, the source of wealth production, the basis for reorganization of legislative and regulatory bodies, as well as changes in governmental methods and the way democracy itself works. In practice, this information society can be expressed as a series of very concrete disruptions involving the post-industrial landscape of the developed world.

The liberalization of telecommunications, the spectacular development of the Internet, and the multiplication of company mergers in IT, media, and telecommunications are, for example, revelations of a single phenomenon, the advent of the information society. By its very nature, the information society ignores traditional boundaries and limits.

Yet, the GI-soc is the product of a much bigger phenomenon than a simple expansion of related interests. In fact, over whole decades of history we ignored, even forgot (consciously or not), about the fundamental role that geography has played in human society. Today, that tendency is reversing itself significantly. For a long time, human societies paid little attention to the spatial, geographic arrangement of people, phenomena, products, etc., and we did not even realize it. Today, though, never have so many maps been produced. Information society as a whole is becoming geographic.

Several arguments can be made to justify and clarify a position that may seem somewhat radical. Some of these arguments can be initially linked to the development of the GI sector, as well as to the diversification of its fields of application. We have been witnessing a considerable expansion in the use of GI in the last decade. This expansion incorporates all spatial and social levels. It involves all types of public and private organizations, as well as the "general public" (e.g. people consulting dynamic, digital maps over the Internet, deliberate generation of geographic content, etc.). However, this phenomenon resonates on at least two other levels beyond the field of GI.

In more basic terms, we are witnessing a return in full force of environmental concerns, at the fullest extent of the term. The state of our planet is one of the major concerns of our modern societies (even if concrete actions are often put off. In 1992, during the United Nations conference on environment and development that took place in Rio de Janeiro, Brazil, Agenda 21 officially recognized the fundamental role of information (GI in particular) in the improvement of decision-making as involves social, economic, and geographic matters. This is in recognition of the need to bring back to the forefront geographic concepts (especially location).

In addition, the development, and especially the exponential escalation of the use of IT ("smart" phones, GPS, personal digital assistants, etc.) creates problems in distribution (manpower, products, services, etc.). The development of an information society comes with the democratization of IT through mobile and recreational uses. During the last few years, this process of democratization has accelerated with the development of Geoweb 2.0 and mass practices in positional technology and GPS navigation. The effect of virtual maps such as Google Earth is real. New technologies and new formats have become more popular (Application Programming Interface, Keyhole Markup Language, etc.); new practices have emerged (user-generated content, deliberate production of geographic GI, crowdsourcing, cartographic mash-ups, geotagging, geocaching, etc.); and new practice communities have emerged (neogeographers). This phenomenon demonstrates the diversity of products and service providers with geographic content, with potential users multiplying and their geographic location forcibly sought. An increasingly diversified number of private companies use positioning data to support tactical or strategic decisions (e.g. the positioning of goods at Wal-Mart branches through Radio Frenquency Identification technologies, or the use of real-time determinations of fleet locations for transportation companies), or offer such services to other companies (spatiotemporal analysis of new markets, positioning of new stores, Application Programming Interface, Keyhole Markup Language, and Radio Frequency Identification).

The development of the GI-soc is, in any case, a recent phenomenon and one that is altogether difficult to define. What are its characteristics? What are its similarities and differences compared with the information society in general? What are its key challenges? What is the contribution of GITs, GISs and, more generally, GI for individuals, organizations, and society? How far will states and governments go to reach their citizens through the help of participating GISs (public participation GIS or PPGIS)? What forms of social interactions are being created among GIS practices, individuals, and organizations? At this point, there are still more questions than answers. However, these questions can form the structure through which the organizational aspects of GISs are analyzed.

This book is based on the simple acknowledgement of our lack of knowledge of this "geographic information society". The difficulty in understanding its challenges is based on three aspects in particular: (1) the relative newness of the information age (as a concept, as social reality, as well as an object of study); (2) the relatively small number of research studies involving this specific subject; and, finally, (3) the problem with understanding the subject because of its "multi-disciplinary", eclectic nature.

These questions suppose that knowledge in fields as varied as sociology, administrative sciences, IT, geography, and geomatic sciences, among others, can be brought together with different cultural points of view. To help us understand the organizational and social dimensions of GISs, this book will propose different angles of attack, summarized by the following diagram.

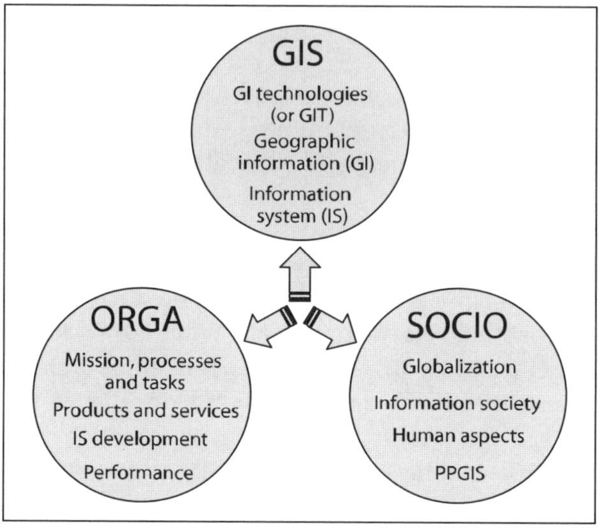

Figure 1. *Overview of the basic concepts of the book*

This figure illustrates the fundamental interactions among three systemic levels: (1) GISs (geographic information systems, in the overall, broad sense of the social and technical aspects of the system); (2) organizations (private and public, more or less structured as associations or groups of citizens); and (3) societal systems (incorporating the human aspects of the system). Note though, that this distribution is entirely imaginary. In practice, distinctions between these systems are not always well-defined. But the identification and consideration of these three levels and especially of their interrelations will help us to effectively cover the questions asked in this book. This figure guided the authors in the choice of which subjects to cover, as well as in the authors we would invite to write on these subjects. This approach made it possible, from the very first, to structure the chapters of the book in the most consistent and balanced way possible.

This book is meant to address anyone with a basic knowledge on GIS design, implementation, and use. It proposes a deeper understanding of the less technological dimensions of GIS development. Part I sets the stage with a global and historical perspective (Chapter 1) of the evolution of the GI-soc from an American/English point of view. It also constitutes a report on the changes that have occurred in business brought about through the development of geomatics (Chapter 2) and builds an explicit link between IT in general and GI in particular (Chapter 3). Part 2 gives the reader a better understanding of the challenges and problems with the implementation, diffusion, and use of GISs in organizations. More specifically, the organizational and social roles and values of GISs are covered in Chapter 4, whereas Chapter 5 provides a report on the adoption and uses of GITs in organizations. The matters of using GIS as a project control method are addressed in their organizational dimension in Chapter 6. Chapter 7 concludes Part 2 with an analysis of the approaches that businesses have developed to use GIS. Finally, Part 3 invites the reader to explore the various organizational and social aspects of the implementation and use of GISs, providing a thematic dialog in which the reader can participate. A case report of a few provincial government agencies in Canada that use GIS is the subject developed in Chapter 8. The use of GIS by local communities in Britain are analyzed in Chapter 9, followed by communities and citizens' associations in the United States in Chapter 10. Finally Chapter 11 concludes with a discussion of inter-organizational challenges in GIS with the help of French and American-English case reports.

We believe this collection can be the foundation for reflections that are necessary to better understand the various dimensions and challenges of the GI-soc. It is vital that this subject, as complex as it is, is understood if we wish to build the future of GIS in a rational manner. We believe that this book constitutes a forward step in the necessary evolution of our knowledge.

PART 1

Issues and Prospects of GIS Relationships, Organizations and Society

Chapter 1

GIS: A Historical Process. The Past as Support for the Future

1.1. Introduction

During the last 40 years, geographic information systems (GISs) have emerged from a complex interaction of stakeholders. As a key sector, the community of researchers increasingly became aware of the scope of the network involved. At first, researchers all strictly focused on the technical aspects. When they addressed organizations, there were obstacles to overcome. When they focused on social interests, it was far into the future, like a final goal to reach, not as a constant source of interaction. More recently, though, studies on interactions among societies, organizations and GIS technologies have become more nuanced.

This chapter addresses the development of GIS from a historical perspective through reflections mainly from English-speaking researchers. We will specifically discuss the point of view of one of the members of this community, whose ideas and theoretical concepts evolved during this period. It makes use of the first person when relating situations in which the author played a direct role.

Chapter written by Nicholas CHRISMAN.

This chapter is biased toward a symmetrist view of technology. The resulting approach is not an attempt to determine which stakeholders are the most powerful or important, but instead it focuses particularly on history and geography. In fact, a similar model does not necessarily lead to identical results when contextual elements change. This incidental point of view is clearly in contrast with the different ideologies of technological determinism, which are supported by GIS technophiles and most GIS critics. The goal of this chapter is to present a study of interactions between humans and organizations in the management practices of knowledge on the environment.

1.2. The original concept

The linear reasoning of the origins and the heroic inventors presumes that a single event is identified as the original invention. However, and inexorably, any invention has its own origins. It seems unwise, even impossible, to determine an origin without considering all the events involved. The current aphorism in this situation can be summarized as follows: "Which came first, the chicken or the egg?" In reality, this reasoning hides many lateral connections and complex interactions. The great idea of a single origin fades in a culinary metaphorical mix, right in the middle of chickens, eggs and spices. There are too many cooks and the logic for determining who arrived first does not really exist.

Nevertheless, this dilemma will not be resolved until a theoretical basis is established. The field of science and technology studies (STS) has long analyzed the overlapping between societies and technologies. Studies such as those from MacKenzie on the precision of rockets [MAK 90] clearly show that technological determinism alone is not sufficient. The progressive improvement of American intercontinental rockets is the product of efforts by stakeholders wanting to reach this goal and not by gyroscope technology guiding the rockets. In parallel, the Soviets chose quite different gyroscope models. The pressures to reach this precision did not exist in France or in China because of differences in the strategy and finances of these two countries. The most significant proof is that civil navigation (which used identical gyroscopes) was not as concerned with greater precision as it was with cost control and reliability improvement. MacKenzie added in conclusion that we could not ignore political pressures in technological development.

The symmetristic approach was developed in STS by researchers such as Bruno Latour [LAT 86, LAT 90, LAT 93, LAT 96], Michel Callon [CAL 81], Karin Knorr-Cetina [KNO 83] and many others [BIJ 87]. This trend came about from a paradigm conflict with a group dedicated to "scientific knowledge sociology" [BAR 74, BLO 76, COL 75], also known as the "strong program". It maintains that social (and politico-economical) dimensions are at the heart of all scientific questions. Symmetrists, however, maintain that the strong program comes from a natural reaction to the previous approach, which preferred studies on scientists to those involving science. For them, it is vital to avoid any separation between social and technical explanations at the risk of rendering the latter vague or even useless.

One of the most widely used arguments from GIS critics involves their origin. Neil Smith [SMI 92] focuses on military roots (as does John Cloud), while others [PIC 95] [SHE 95] connect to modernist philosophies going right back to the Enlightenment. In a way, all these arguments are relevant; they are all "correct". However, the idea that a single origin is the basis of GIS development must be refuted.

1.2.1. *Origins of GIS*

If we mobilize the concept of "rhizome" developed by Deleuze [DEL 76] to analyze the technology, the linear logic involved in the current significance of "origins" is questioned. In general, GIS technologists believe in historical stories about the heroes in a mythical past, who created GIS with a single vision. In the most recent reports, Roger Tomlinson (geographer) and Ian McHarg (landscape architect) appear as the creators. The book edited by Foresman [FOR 97] uses a North American "border" metaphor involving avant-garde "pioneers" who conquered new territory to be established. The reports on origin claim paternity for the "first GIS" [TOM 98] and demand recognition and particular status for the creators. Roger Tomlinson has certainly made a significant contribution to the design of the *Canada Geographic Information System* (CGIS) before its publication in 1968 [TOM 68]. We in no way wish to minimize his contributions. Nevertheless, CGIS did not come from nowhere.

In a retrospective publication, Tomlinson [TOM 98] maintains that GIS has its origins in the public sector supported by commercial consultants. If we limit ourselves to the question of the first use of the term *geographic information system*, he is totally correct, but if we address the origins of specific software and model collections enabling our current GIS to operate, then the origins are a much more complex problem. The program written by IBM for CGIS was not installed anywhere else. Only two CGIS scanners were built: the original one and the one installed at IBM offices in Ottawa to ensure the maintenance of the original. The most revealing proof of a single "origin" is often supported by the community in which it is seen. Tomlinson made his first presentation [TOM 68] during a conference in Melbourne, Australia, for real-estate professionals. These professionals quickly understood the case developed by Tomlinson insofar as he supported jobs that they were already responsible for, but with different technologies. That same year, Samuel Arms presented a review of four "GIS" software packages at the sixth annual URISA meeting [ARM 68]. If we find four alternatives during this URISA conference, can we still consider that CGIS was at the root of GIS? In fact, even though it was the first, CGIS leaves no direct tracks today. Tomlinson was right on one point: CGIS is a vital reference, in particular, in the organization of some of the very first conferences on GISs, as well as from working as a consultant on cutting-edge applications.

Tomlinson [TOM 98] recently attributed purely Canadian roots to CGIS. He was sure that the personnel and funding were Canadian. As for the rest, it is not certain that technical solutions (such as the Morton index or the calculation of a polygon surface) came from the policies or the culture of Canada. CGIS emerged from a group of practices combining natural resource professionals from all English speaking countries and particularly the colonial faction. Tomlinson mentions the contacts he made with Spartan Aerial Surveys during a flight to Nairobi. Natural resource experts have had to create surface measures with planimeters or dotted grids for decades. Computers helped with these tasks by offering different working conditions. These changes were generally technical, but very early on, the justification involved economical, labor, and expertise value issues.

Tracing the origins of GIS requires more than just following the life of the great historical figures. Software codes were also very useful in revealing how the past continues to influence the present. An exploration of the two major technologies, raster and vector, is helpful in demonstrating how the rhizomes are more useful than the dendritic root model.

1.2.2. *Raster origins?*

Whether they are raster or vector, all GIS programs have origins in the Harvard Laboratory for Computer Graphics and Spatial Analysis. The "Spatial Analyst" extension of the ArcView program from ESRI is only one of several implementations of Tomlin's algebra and his MAP Package. Tomlin first created these functions for IMGID, developed by Sinton, when he was a student and was working on a project for Steinitz involving urban growth. Similarly, ERDAS, which was founded by two other students of the same environment, named its first program IMGRID before it changed to IMAGINE. These connections seem to connect Harvard University to the commercial world in the mid 1970s. However, it is possible to find older connections. For example, studies by Steinitz correspond to the superposition analyses carried out by McHarg and other landscape architects before him [STE 76]. In current GIS applications, it is easy to find cartographical overlay functions ([LAP 97], for example) used to define the stacked cartographical layers such as "cost-grids". The use of these surfaces representing the cost function relates to approximation by minimum cost increase methods developed by Tomlin to minimize the movements on these surfaces [TOM 90].

However, where does this idea come from? This component of GIS technology is not present in the inventory of natural resources listed by Tomlinson in the 1960s. Nor does it come from McHarg's superposition of transparent layers method. We can, however, find origins in the studies by William Warntz, second director of the Harvard Laboratory, focusing on the theory of surfaces and particularly minimum cost geodesy [WAR 57, WAR 65] [WAR 66]. Warntz was trained, as several other geographers of his era, as a navigator on Second World War bombers. These planes did not reach the stratosphere. Navigation had to rely on atmospheric pressure curves. Warntz learned to analyze surfaces from weather charts and applied it to his own theory of surfaces. In the 1960s, his work was purely theoretical. There was no computer development. Nevertheless, the analysis of surfaces was part of education. It influenced future programmers who incorporated it to software when computers were powerful enough to support large-scale analyses. The role of Harvard University is clearly not vital. Other geographers received the same navigation training, but few of them had the opportunity to have a real influence during the creation of commercial software. Warntz was influenced by Stewart's "social physics" and by his military service. The simple term "cost-grid" opens up a large network of related origins.

1.2.3. *Vector origins?*

In the vector area, some specific characteristics of current GIS software come from other Harvard University teams. The ODYSSEY project [DOU 80, CHR 92], for example, made it possible to develop superposition functions of complex geometries with "fuzzy tolerance" to avoid parasites and other geometric problems, which inconvenienced prior systems such as CGIS [GOO 78]. ODYSSEY specifications were then provided to ESRI through Scott Morehouse (member of the ODYSSEY team). They were then implemented by a team of geographers at SUNY-Buffalo (these geographers abundantly describe the studies by the Harvard Laboratory). The product of this development was named Arc/INFO. It replaced the non-topological PIOS system. ESRI competitors had to update by incorporating identical characteristics, to be able to gain some market share. With the help of GIS software analysis, we can find several possible traces of their origins. However, these traces are connected to people, education, ideas and organizations. The network is not as linear is it appears. ODYSSEY project was created at Harvard University in 1979 in full disclosure of the studies carried out in the United States and Canada. The topological concepts of ODYSSEY owe more to American census board than to CGIS. Even though Harvard University contributed to the study of American census during the whole of software development period, the topology was not immediately taken into consideration. The problems with the software were quickly used as examples to demonstrate the importance and relevance of the topological model [PEU 75].

The history of the Harvard Laboratory for Computer Graphics and Spatial Analysis actually resembles a labyrinth. There are a few precious opportunities for analysis or observation. It does not involve the linear recall of GIS roots, but instead a series of interactions within a community with common interests. The rhizome–Deleuze–metaphor applies here. CGIS, Harvard University Laboratory or any other GIS do not have a single origin. Howard Fisher, the founder of Harvard University Laboratory, first heard of digital mapping during a workshop led by Horwood in 1961 at Washington University [CHR 97]. Does this mean that all laboratory ideas come from Seattle? Of course not. As with heraldic lily rhizomes, the world of technology has multiple convoluted connections, with or without conclusion. However, sometimes a few flowers emerge on their own through this maze of connections.

1.3. Geography discovers GISs

In the early 1990s, geography (specifically as a university discipline) became aware of the appearance of GIS. A sub-discipline then appeared in multiple forms. The Association of American Geographers (AAG), for example, created a GIS-specialized workgroup, which expanded over time to become the largest group in the organization. A large number of geographers decided to focus their efforts on the growth of GIS. The attraction was not the novelty since the subjects of GIS research started 30 years ago; there were also many more recent research studies. The attraction most probably came from the strong market growth instilled by the GIS industry and indirectly from users. This comment is not meant as another celebration of the exceptional growth of GIS. However, this increase must be taken into consideration, as it is an integral part of GIS history.

The recurring criticism and self criticism in social sciences concerning cartography began to take shape [HAR 89] [HAR 90] [WOO 92], sometimes with a clearly anti-technology message toward those who blindly believed in GIS. For Harley and Wood, knowledge of maps meant power: the imperial power linking politics, the economy and society serving a fortunate few. The interaction between knowledge and power goes back to major social theories, such as the one from Foucault with a panoptic character [FOU 77], or by more abstract generalizations from "Seeing like a State" [SCO 98]. While this debate was taking place on cartography, the attention moved toward the GIS phenomenon. During a calmer period, a publication [DOB 83] praising "automated geography" (one of several expressions to describe GIS) received a lot of attention. However, most comments came from GIS specialists. During the publication's 10th anniversary, a retrospective [DOB 93] garnered comments from much more varied sources this time. Other times, goegraphers also questioned the GIS phenomenon [CUR 91, SMI 92]. However, the cases of excessive "defense" [OPE 91] (among others) were those attracting most of the attention [TAY 91].

This period of misunderstandings and excessive attacks has fortunately ended. The publication of *Ground Truth: The Social Implications of Geographic Information Systems* [PIC 95] modified the dialog insofar as the book gave voice to GIS experts (particularly Michael Goodchild [GOO 95]) and to more critical opinions. Even though the title *Ground Truth* (a very nuanced choice of words invoking the remote sensing practice of references on the ground, as well as a sense of the "grounded theory" social critique type) is a good seller, the often ignored sub-title imposes a direction: the implications and consequences of technology. This book constitutes an important testimony on questions raised by the GIS-society relationships. It also led to the creation of an ad hoc university community. During the time this movement was in contact with the National Center for Geographic Information and Analysis (NCGIA), a research program structure was taking shape. After a meeting at Friday Harbor in Washington, Eric Sheppard [SHE 95] presented a research program on GIS implications. The first chapter involved the history of GIS development from an intellectual standpoint, focusing on the environment in which its tools were created and detailing the social results of certain choices. Sheppard chose the expression "social construction" even though his theoretical basis was more in line with Habermas and Heidegger who are considered idealists rather than constructivists. The second and third chapters were completely dedicated to the impacts and implications of GIS. This research program, through NCGIA procedures, became the central element of Initiative 19 [HAR 96] called *GIS and Society: The Social Implications of How People, Space, and Environment are Represented in GIS*. Despite some minor attempts [CHR 96] to direct reflections toward the analysis of reciprocal interactions between GIS and Societies, the "GIS-Society" body (now a sub-sub-specialization) concentrated all its efforts on questions of impacts and implications. The social dimension of Project Varenius, for example, focused exclusively on GIS implications on society [GOO 95a]. Similarly, the research project written in 1996 for the University Consortium for Geographic Information Science (UCGIS) focused exclusively on these implications [UCG 96].

However, the discussions involving axes of research, even as elaborate as they are, cannot replace the research itself. This chapter actually emphasizes how important it is to expand the research axes by avoiding one-sided approaches in the field of "GIS and society". The strategy developed here involves the consideration of all the impacts and implications to close the loop. "The social context influences the development of GIS as this development modifies society" [CHR 96, p. D9]. Closing the loop makes it possible to consider the theoretical problem with GIS much more efficiently than theoretical back and forth [PIC 97], which is too often divergent.

1.3.1. *The scientific debate ignores the main question*

GIS supporters very often present geo-computer technologies as tools for more efficient and fair decision making. They also emphasize the support of these technologies to remove questions that are too subjective and to rationalize the creation of a consensus, so that decisions may be more objective [COW 88, DOB 83, DOB 93, MOR 94, OPE 91]. The major part of these studies is in line with the "Course of Progress" model, a vision of history limiting the possibility of working out choices and inconsistencies in technological transformations [CHR 93]. Following this model, the idea of automated geography implies that technology is independent of people and that it operates under its own internal logic.

The search for optimization of objectives or decisions with the help of mathematical methods (quantitative) is not a recent occurrence. GISs do not infringe upon the computer frenzy of the 20th century. They are not isolated cases. Porter [POR 95], for example, describes in detail "trust in numbers" inspiring the stakeholders involved in the construction of canals in France, in the choices of railroad locations or in calculations of cost and performance of dam projects from American engineers. The GIS is only one in a series of practices established in deep historical systems. However, the geographic differentiation of information mobilized in GISs brings an additional level of complexity.

Discussions of GIS technologies often end in a speech full of technological determinism. GIS supporters and critics maintain (consciously or not) that technology is intrinsically independent of the social world. This led to the two major principles of technological determinism [BIJ 87, BIJ 92, FEE 95, WOO 87]:

– Technological development follows unilinear progress of less advanced systems toward more sophisticated systems.

– The technology constitutes a necessity to which social institutions and physical persons must adapt.

This technological determinism implies that technology can be studied apart from the context in which it was developed or used. The only thing left to study is the "implications". I therefore propose to study the way in which GIS technologies are used in practice, reinserted in their social context, instead of their own.

1.3.1.1. *The myth of progress*

The simplicity of a linear progress path dominates auto-representations of this new GIS industry [DAN 84, TOM 84, TOM 89, ANT 91, MAG 01]. The heroes of progress imply that the improvement is inevitable and assured once and for all. Adherence to GIS would guarantee success. GIS technology would satisfy all demands. For Dobson [DOB 83, DOB 93], the GIS technology is the next logical step for a better future: "GISs have become the *sine qua non* condition for the geographical analysis and research... they define the first phase of a technological, scientific and intellectual revolution" [DOB 93, p. 431]. The authors of *Ground Truth* put a lot of emphasis on claims from GIS supporters [PIC 95] and sales professionals [GOS 95, ROB 95]. The more arrogant the claims, the more they serve the critics. Pickles, for example, describes his unease [PIC 97, p. 368] concerning the commercial success of GIS. These attitudes and positions can be interpreted as attempts to find support. GIS supporters see the proof of the legitimacy of their claims in their success. Critics judge this success as disgraceful for a university discipline. Each will interpret the same events with their own perceptions.

The main school of GIS methods (called GISciences) claims the abstract purity of geometry and mathematics as a base for GIS tools [GOO 87, GOO 92]. It is a perspective derived from abstract essentialism (for illustrating its most extreme form): the geometric concepts are located in a platonic world of forms; researchers find these eternal truths; technology can rely on the iterative approximation to achieve ideal truths. In this context, social interests are only secondary since the only way that society can block progress is to delay the inevitable. The GISciences coalition [MON 95, UCG 98] firmly clings to technological studies involving cognitive, not social, dimensions (see [HUT 95, WOO 87] for a critique). Nevertheless, some research studies are beginning to link the social context with technical implementation [CAM 93, CHR 87, CHR 91a, CHR 92, HAR 98, HAR 00], thus positioning themselves against the GIS community resisting technological determinism.

1.3.1.2. *The response from critics*

As mentioned previously, GIS critics [CUR 91, PIC 95, SHE 95, SMI 92] focus on the impact of the technology. This is the reason why they often present technology as an autonomous entity, without social control and external to political speeches. It is basically a somewhat sophisticated version of the "two cultures" model of Snow [SNO 59], which shows that technologists do not read the same literature and are not committed to the same philosophical speeches (Pickles [PIC 97] specifically emphasizes this division). Even though there may be theoretical divides, technology and technologists should nevertheless always be able to respond to social pressures. Not reading the same material does not erase the fact that they live in the same world.

Technological determinism, openly proclaimed by GIS supporters or implicit with critics, overshadows relations between GIS technology and society by ignoring certain links. The controversy between progress supporters and critics armed with their social theories ignores the people involved in the technology and the different social interactions behind it. GIS technology expands human capacities in other ways. GIS users are no more the instruments of progress for information technology improvement than the victims of resulting social consequences. Current systems reflect the many layers of negotiations between objectives (social) and capability (technical and other) to respond to them. Simplistic metaphors must leave room for a more nuanced understanding of interactions between man and technology.

GIS technology is the result of localized social developments and not the result of a vast superhuman kingdom. These developments occur when technology is created and continue as long as it is configured for each application. These developments were not created overnight, but in a continuous way over a certain period. The myth of progress must be replaced by a scrupulous examination of the development of geographical computer technologies in their specific context of time and location. No historical or ideological schema alone can explain the complex interactions of a specific place, or the variation between distinct locations. The development of a GIS in a county, for example, in the United States, involves combinations of ideas as diverse as those of concepts, people and corporations from ancient Rome, medieval property schemas, concepts of equality from the Century of lights or more recent concepts of environmental responsibility [CHR 93]. The geographical computer technology hence supports several concepts simultaneously, including the possibility that conflicting ideas may co-exist within a single institution [LAT 93, USH 54]. The spatial and temporal mix of ideologies is obvious in practice, but the abstract essentialism, which dominates research circles, too often ignores it. Critics also refuse to face these mixes insofar as the major part of their argument is based on the metaphor of the original sin, explaining everything from single origin. In addition, the use of a temporal rhetoric represents a powerful force in the development of sciences and technologies. Associating work with "progress" implies committing to "irreversible" action [LAT 93 p. 51]. Just as reports on origin are based on the past, the myth of progress relies on the future, so that nothing can interfere with the expected configuration. In paradox, GIS research involves timeless universal ideals at the same time as they attempt to reach some level of progress.

Society in general, as well as a variety of social structures, influences the nature of geographical computer representations. Reciprocally, some of the geographic information characteristics influence society. Research of GIS only rarely considers the gray zones involved in this dual meaning. The challenge for researchers in the "GIS and Society" sphere is to close the loop of implications and interactions.

At first, research in terms of GIS was carried out with no consideration for social factors. To date, GIS principles are presented as universal constants, resulting from abstract geometric laws. Social, cultural and historical contingencies in the development and use of GISs are considered aberrations, which distorts the logical path of technology. Some use it as a model of a "barrier", explaining that natural growth and distribution are hindered by irrational (and unexplained) social forces [CRO 91, ONS 91]. Technological innovation is not a static force, but it is the product of complex interactions among economical, institutional, political and cultural stakeholders.

Some pragmatic research projects are conscious of their specific origin. They admit that the technological solution developed responds to very specific societal demands, that it is not the product of a universal course of technological progress. That is, for example, the case with the Dane County Land Records Project [NIE 87]. In fact, several models of soil erosion exist in different and diverse scientific disciplines. The American federal government demanded these models be applied throughout the whole territory. After long negotiations that were more political than scientific, it was agreed that GIS technology would provide ways to respond to this request. This type of example is not adapted to the course of progress formula because technology brings a specific answer adapted to a specific societal request. The direction then goes from society toward the technology solution and not vice versa. Society as seen by pragmatists has great ambitions. This was the case, for example, of very sophisticated large-scale environmental plans. However, these plans were too expensive to be implemented. Information technology then made it possible to bring to the forefront dormant ideas. These plans are basically the product of a tortuous series of rhizomes, giving origins of the difficult task of following the necessary paths of dendritic structure. Sometimes, technological development takes credit for the history of the Gordian knot: technology cuts across complex labyrinths to reach the goal in a clearer and more direct way. It is another way to find a control point.

1.4. Trying to find assistance: the study of sciences and technologies

Contrary to technological determinism, which prevails in most GIS research projects, this chapter relies on recent theoretical bases from crossing sub-disciplines such as science sociology, history of technology and sciences or science philosophy. In the beginning of the 20th century, we saw marked interest in the cumulative expansion of sciences [CAR 66]. Halfway through the process, rational positivists seemed to have overcome all opposition, advocating the method as a central path to scientific coherence. Kuhn [KUH 70] demonstrated that science is not as linear as that. The development of relativist theories in physics, for example, required the adjustment of the complete "paradigm" and not only the incremental accumulation of adjustments of some established schemas. Kuhn's approach maintains the independence of science (and therefore technology) from social interests. His ideas were predominant. To such a degree that quantitative geography adopted the paradigm terminology ([BER 73], for example) fundamentally based on an anti-positivist theory of knowledge. Some recent studies in the history of sciences [GAL 97] reveal new improvements in terms of scientific process understanding, expanding the concepts of paradigms to include negotiations between theorists and instrumentalists. The certainty that a specific scientific method still works is in question [FEY 93]. History and philosophy of sciences no longer support the inevitable progress myth.

Several studies in sciences and technology ([BAR 74, BLO 76, LAT 86], for example) provide important documentation on complex networks linking social organization, political structure, economical interaction and cultural foundations of the development of a technology. Sociology of scientific knowledge developed an important "program" [BLO 76, COL 81]. It maintained that social relations are the direct result of technological or scientific actions. From this point of view, the study of impacts on society was ignored insofar as we go from society to sciences. This position is orthogonal to the direction taken in "GIS and Society" research programs.

Writings from constructivists [BIJ 87, BIJ 92, CAL 81, KNO 83, LAT 86, LAT 87, LAT 93], although very different and not coherent, nevertheless reject the idea that the dynamics of mutual constitution work only one way. The French school of science studies is called "*école de traduction* (school of translation)", but this term is not applied in English. Latour ([LAT 93], for example) maintains that the division between nature (undeveloped but discovered scientific investigation zone) and society (zone developed by man) overshadows the tricky interactions required to maintain hybrid networks (mix of nature and society) constituting technology.

For constructivists, science and technology are developed from several points of view, and the development is clearly local, never universal. Several social forces work together to form complex technologies such as GIS. The development of any technology depends on a specific environment, which is an important source of constraints in terms of the technological authority of the stakeholders involved. After a study of the first geophysics steps carried out by Schlumberger, Bowker [BOW 95] reports that the results exceed the idea of simple rooting in a specific time and place and are closer to a series of selective interactions which influence technology. These writings reject the idea of a rationality inherent to technology. Bowker thinks that any logic present in a technology was placed there by developers to respond to a certain number of requirements, but that it is adopted by users for their own personal reasons.

It is becoming increasingly difficult to separate the technology of sciences, which is very positive for GIS, because of the connection between the tools and disciplines using it. The constructivist approaches provide a theoretical framework to explain the tight links between actions taken by people and the technology they develop and use. The network built by technology and society includes complex relationships between artifacts and people, between institutions and data and between software and researchers. Mack [MAC 90] describes the interactions leading to the Landsat system as a result in no way determined by relentless forces. Martin [MAR 00] used the method of networks-stakeholders to explain how GIS institutions behave in the equator. For a long time, GIS researchers observed how institutional factors play a major role in the success or failure of a system. However, we have very few references enabling us to voice any predictions. Network stability makes it possible to verify the viability of an organization. However, this is only a descriptive approach and not a predictive one.

The mediation of different interests occurs during the development of artifacts; negotiations are necessary for corporate consistency. Star and Greisemer [STA 89] and Fujimura [FUJ 92] describe the relationships between the many stakeholders and their artifacts with the term "boundary objects". These are objects acting as mediators between the different groups, without creating a consensus that is shared by all. On the other hand, they offer a mixed function: they distinguish the differences between the groups and provide common reference points [HAR 98]. The groups agree on the definition of the boundary object in different ways. Organizations, in the form of institutions or disciplines, then play a central role in the production of translations, which are stable between each point of view. Galison [GAL 97] goes deeper in the exploration of this boundary concept and gets closer to the multi-disciplinary practice of GIS. Instead of a translation implying mutual understanding, he uses the linguistic metaphor of a Creole (or "pidgin") dialect limited to a specific zone of exchange. In terms of GIS, a common language has become a recurring dream, but dreams must be abandoned, and we must face the obvious that real progress in sciences only occurs with partial agreements and incomplete participation.

In the field of GIS, any data sharing agreement creates boundary objects. They can be physical structures, concepts or normalized processes. In any case, translations are incomplete. The multi-use cadastre layers serve as examples: each skill is responsible for its own layer, but a system of reference ensures some compatibility with the others. Each GIS contains a unique collection of artifacts enabling multiple social groups, with different, or even contradictory, values to find compromises to develop other technological artifacts to share. In the best case scenario, the social development of technical objects is only stable for a specific moment and it is a constant subject of renegotiations.

1.5. A few fundamental principles

In 1987, during the 8th AUTO-CARTO conference, I presented a paper called "*Fundamental Principles of GIS*" (then published in *Photogrammetric Engineering and Remote Sensing* with a more descriptive although less striking title: "*Design of GIS based for social and cultural goals*" [CHR 87]). I do not wish to wax nostalgic here, but to use this writing as a loop for examining the state of GIS research and the progress of the understanding of organizational questions and societal connections.

Returning to this period to set the scenery is not useful. The GIS body then continued to be amazed by its quick expansion. In the Fall of 1986, the British organized an AUTO-CARTO conference in London. This conference was very successful. In its wake, the first GIS manual [BUR 86] was published. In retrospect, in the second part of the 1990s, the GIS community tried to impose topological principles to structure the software. Intergraph, for example, developed TIGRIS [HER 87], an object-oriented system using a topological structure indexed with "R-trees". Even though this approach corresponded to the presentation I made during the AUTO-CARTO 2 conference [CHR 75], I had already revised my concepts toward more decentralized models.

The main argument of the report presented at AUTO-CARTO 8 (and the most quoted since) was that GIS technology "must be economically, as well as politically, socially and even ethically responsible" [CHR 87 p. 1367]. This type of argument seems indisputable today. But then, the idea was quite innovative or even provocative for those who adhered strictly to the technical field. My report focused on the design and development of databases. In the preamble, the raster/vector controversy was alluded to as a closed subject (without much apology for my personal role). I was attempting to highlight some downsides to the trust placed in requirement studies for users because these studies had a tendency to confirm the existing status quo. I then presented the traditional communication model as the source of the problem and proposed a sensible alternative to cultural and historical questions. It is highly probable that very few readers could understand the diagram I created to explain how data storage in institutions is the historical result of interactions between people and their environment. When I reexamined this diagram in 1992, I found it too structural [CHR 92]. The report also addressed the problem with geographic information objectivity, which today may seem somewhat obsolete. The development of these ideas remains active, and it is only in the last few years that I started to find resources in STS.

The positive contribution of this report was in its legal approach of the needs analysis based on the mandates assigned to the stakeholders. In a way, the coordination of efforts in the GIS field confirms the relevance of this approach in the last 16 years. It is not surprising that Wisconsin or even the US Federal Geographic Data Committee, for example, among many others, have adopted this strategy. However, in most cases, the process involved follows the reverse route. Instead of rethinking the contents of databases right at the beginning using the real mandates of stakeholders as guides, it seems that most organizations impose their views by political processes and power struggles, in the development of databases, which actually confirms divisions of labor as they exist without necessarily considering the mandates and actual roles of the different users. In all cases, the result is more decentralized and cooperative, compared with alternatives put forward at the time ([NRC 83], for example). My voice was one of many, and the idea of cooperation assembled a large number of supporters. In retrospect, centralized databases were already obsolete in 1987, but the conceptual and technological improvements could not be easily seen because of the lack of power from the computers back then. It was years before these movements could find some form of support from the GIS community.

In conclusion, my report supported fairness, and not the efficiency measures that were predominant up to that point, during the evaluation of GIS. "Geographic information systems should be developed on the primary principle that they will ensure fairer treatment of those affected by the use of the information" [CHR 87 p. 1370]. I also used the occasion to denounce the "raster" approach by arguing that GISs must retain the identity of the interesting elements, instead of imposing arbitrary objects. I finished with a reflective criticism, saying that the report presented during the AUTO-CARTO 2 conference [CHR 75], which proposed a database that would be topologically integrated "was flawed because it centralizes definitions" [CHR 87 p. 1370]. This GIS development approach proposed moving the authority too often retained by a technological elite, in order for it to be closer to political and administrative considerations. The last sentence introduced the following debate [CHR 87 p. 1370]:

> *... the search for technical efficiency must not be allowed to overturn political choices without careful examination through the political process. The true challenge is to use the increased sophistication of our automated systems to promote equity and other social ends which will never fit into a benefit/cost reckoning. I am convinced that the future of GISs will lie in placing our technical concerns in their proper place, as serious issues worthy of careful attention. These technical concerns must remain secondary to the social goals that they serve.*

1.6. Closing the loop

In most discussions, GISs are presented as an integral unit. It is true that a "system" can be as inclusive as we want it to be. Nevertheless, to efficiently trace the paths of influence, separating GIS components is preferable. There are several ways to separate a GIS into sub-systems. However, we must realize that, in addition to the "software" component, the description of "data" and "results" components makes it possible to cover and understand most paths of influence in which society influences GIS, even as these GISs influence society in turn. In any case, each of these components reveals a specific perspective in the way GISs, institutions and society interact.

1.6.1. *The "software" component*

In most critiques against GIS, the software is considered a single force, not an object painstakingly developed by individuals. In reality, the commercial power and size of some corporations hide the very nature of the GIS software development process. First, GIS software requires very few people and locations throughout the world. Programmers are generally highly qualified, and (at least in the past) tightly connected to the research community involved. These programmers are influenced by the solutions that they have already acquired and the lessons they learned from their previous work. Paradigms in terms of programming follow the same patterns as fashion, they come and go. At some point, the trend was bandwidth scanning algorithms [BEN 80], then it was spatial indexation or object-oriented approaches. Each of these paradigms made some operations easier, but others became more difficult. The artifacts created by programmers became stakeholders themselves. Who has not heard the phrase "Arc/INFO does not allow me to do this or that"? From this point of view, the software marks the limits of what a GIS can do, but these limits are difficult to find. Ideally, the users, and not the programmers, should be the ones to decide what functions they actually need. In practice, users establish combinations that were not originally created by developers. They often push the software to do things that are far from their original specifications, whereas other capabilities are never used (this observations goes back to the concepts of usage technology and espousing technology presented in Chapter 5).

Because of the many levels of abstraction used, programming mobilizes most negotiation metaphors. Several modules share (memory) space by sending and receiving messages, by hiding internal details and by sharing results. This process can be interpreted as a long chain of moving binary elements generated by a Turing machine. Each module has its own idea of the process, which is negotiated by some forms of boundary objects. It is almost impossible to perfect a program without visualizing the social communication world created by the program code. This form of society generally presents fatal problems, unmet expectations and crossed behavioral boundaries. Smart programmers build their world as scaled models of railroad installations in which they can let the trains work independently. In reality, in the 1960s, MIT programmers of the first "hacker" generation, for example, were recruited in large numbers by the "Model Railroad Club". Hackers are often considered as anti-social people. Their abstract world is characterized by heterogenous interactions however; probably, because not everyone can see the trains circulating in these abstract worlds.

The software is characterized by its own historical process . In a way, software is a relatively new field, and reflections on it are just as recent. Decision making involving a storage format or a transfer standard can become an obstacle to maintenance, insofar as it requires to constantly handle misunderstandings. The choice for continuing to only use two digits for the "year" category format involved important investments in terms of work at the end of the last century. There are still some operating systems, although long forgotten, which limit certain evolutions to this day. The more complex the interactions, the less creative the field of operations. In terms of GIS, the old errors have become *de facto* standards, whereas official standards which were painstakingly adopted become obsolete (often for good technical reasons). The angular Babylonian units of measurement are still current because they are deeply rooted in practice. Planar projection systems are still persistent even though computers can calculate in an ellipsoid. Any decision taken at any time can be questioned when cost changes in relation to some system components are considered.

Despite an internal dynamic and idealism linked to computer language abstractions, software meets the needs of users. Systems get developed according to commercial instructions. The design is carefully done, but the less visible parts are ignored. The discovery of GIS arrived at the same stage as requests from potential users. However, this chronological order cannot be technologically determined insofar as it is connected to the economical context of the different sectors. Within large systems, a mistake can be dormant for years and not be detected, and the community can encounter it many times. The commercial side (marketing) tries to "configure the user" [CHR 01], but the behavior of users can also modify the software. Work practices evolve to accommodate new technologies, enabling division of labor and knowledge to be renegotiated.

Despite its apparent independence from social unrest, the world of software is linked to several external interests. Software creates social relations and is, in a way, the result of these same social relations. Till date, however, very few studies have really focused on social connections of GIS software ([CHR 99], for example).

1.6.2. *The "data" component*

The authors of *Ground Truth* emphasize how GIS databases are the expression of political and economical forces. Taylor and Johnston [TAY 95], for example, demonstrate the tight link between wealth and GIS resources. It is not really surprising that such a correlation exists. However, delving deeper into this analysis, which seems even more interesting, is the geography of geographic information [CHR 91]. This makes it possible to better understand the effects of places and contexts. Political and institutional infrastructures of national cartography purely vary from civil organizations, with interest in natural resources or civil engineering, for example, to military organizations motivated by questions of national security. Some countries have very centralized organizations, for example, Sri Lanka, where close to 95% of cartography professionals work for the national cartographical agency. The Sri Lanka mapping agency is a relic of the British colonial administration, still connected to military power. In other countries, the private sector is much more important. Japan for example has more than 940 professionals per 1,000 km^2. The private sector responds to very high demands for land transfers, following the explosion of the speculative period. The maps are then the product of different regimes or ideologies. They are specific and fundamentally different from one context to another and from one period to another. Social objectives thus served are controlled differently. Consequently, these differences express technological distribution in a different way. In addition to simple-problem acknowledgement, researchers interested by GIS and society questions cannot save the cost of a precise analysis of the geography of geographic information.

Goodchild [GOO 95 p. 47] notes that a geomorphologist will be more inclined to choose a representation model of the topography than a computer specialist. This point of view demonstrates his preference for concrete representation. A computer specialist obviously can produce a more efficient model using other metrics for judging topographical representations. The role of disciplines is greater than the variations between countries. A discipline is an institution that organized itself to transmit some form of knowledge and certain values. Beginners easily succumb to indoctrination to have access to the lingo and tools of the trade. The disciplines constitute forms of social technology. Some are more flexible than others. Some mobilize their power to prevent access to resources and to ensure continuity [CHR 92]. Others are more open to intellectual speeches and easily develop new ideas. In fact, the most appropriate metadata for some geographical data is present in the manuals, definitions or scientific citations, which are specific to disciplines. GIS data is not sold by the pound or by the number of coordinate pairs. The professional position of the stakeholders involved in their creation is a founding element (Chapters 2 and 7 clearly demonstrate this). In the early 20th century, topographical maps were signed by the topographer (who followed the map from samples on the field). As the photogrammetric methods of production were changing the work processes to attain greater accuracy, production was becoming anonymous, made up of interchangeable components. However, the wheel keeps turning, and more recently, the standards in terms of metadata come down to notions of individual identity and responsibility. Sources of data available in a specific place make any general software able to work in this specific location with specific people and execute specific tasks.

This makes for an interesting potential for closing the loop between the social context and GIS.

1.6.3. *The "results" component*

The most important is the observation that the software and data do not necessarily determine the results. Given resources can be combined in different ways to serve different and various goals. The simplest tools, even in GIS, can help in obtaining different results. However, the majority of these products make no sense insofar as there is no set rule to protect the user from incorrect or malicious uses. Hopkins [HOP 77] attempted to limit some of the worst excesses in vain. In the end, results must be verified by people using reliable systems. Porter [POR 95], for example, demonstrates very clearly how the increased use of quantitative approaches is considered as a trustworthy technology. This certainly applies not only to accounting [RIA 95], but also to geography. Accounting tries in a formal manner to restrict creativity in order for finances to make sense for everyone. Similarly, maps simplify and formalize the data represented, even if it is not always easy to formulate the rules in writing.

One of the current challenges in terms of reflection on "GIS and Society" involves public participation and access to technology and geographic information. In this way, the political dimension of GIS involves the stakeholders and the results. Concentrating on public participation is a revealing step in an awareness of the importance of the user's identity. Gradually, it clearly appeared that there is no guarantee that the results from GIS can serve a large number of centers of interest, unless of course a larger diversity of stakeholders gets involved. Even though a few initiatives are already taking place on these questions (Chapters 9 and 10 in particular provide an overview), there is still much to do. In summary, we must admit that GIS results are not simply the product of an inevitable technical force.

1.7. A conclusion geared toward the future

The number of researchers involved in the "GIS and Society" problem greatly increased during the last decade. Research dedicated to public participation (PPGIS; see Chapter 10) clearly constitutes the most dynamic reflection trend. A series of annual conferences has actually been dedicated to it since late in the 1990s [OBE 98]. Questions raised earlier in this chapter have somewhat been ignored although they are revived every now and then [SCH 00]. A solution to reactivate them would be to link GIS research to English academia geography trends. However, this option is a contradictory direction to the GISciences movement. In fact, GISciences is growing apart from geography to form its own discipline. Kwan [KWA 2002], for example, wants to "reconsider" GIS as a feminist research tool. Schulman [SCH 02] attempts to reconcile the form of realism predominant with social critics (derived from Sayers) with constructivism of sciences study. This approach actually is closely connected to works from geographers who mobilize the study of sciences in reflections outside of the field of GIS [DEM 96]. Reports on these studies do not provide any more clarity on the retrospective than the approach developed in this chapter could do. However, a few trends are emerging.

This chapter highlights the importance of the concept of control point to efficiently analyze the evolution of a technology. It is not a deterministic interpretation. Currently, it is vital that reflections involved contribute to a democratization of data, to a greater software openness, as well as to the development of ethics in GIS applications. Democratization confronts bureaucratic and commercial forces who are only interested in the bottom line. In Europe, the trends involve data trade, and it seems that costs are less prohibitive. In the United States, on the other hand, where free access was predominant, domestic security priorities have changed the trend. GIS software seems very concentrated; only a few firms completely dominate the market. Perhaps, the arrival of the Open GIS consortium will encourage the development of more open, freely distributed, systems. Ethics come from education. If users can understand that they are responsible (and not the computer or software), maybe then results can be more reliable. In any case, all these possibilities are in the hands of the GIS community. No elite should choose for its best interests, unless, of course, we let it happen.

Most researchers in the GIS community clearly consider the "GIS and Society" problem as a peripheral question, compared with their primary scientific interests. It is obvious that for those interested in these questions, the strategy of considering users as pawns in a military industrial complex (or any force in the current society) is not liable to attract the attention wanted. The odds are that increased interest would emerge if we could clearly demonstrate how institutional and social factors influence scientific centers of interest and the direction of research, for example, the case with ontologies. It is also necessary to highlight the limits of unconsidered refocusing in individual cognitive aspects, which underestimates the role of relations and social networks. However, shifting this paradigm will not be easy, because it is even more deeply rooted than technological determinism discussed earlier. That is why we think it is particularly important to really understand the challenges and practices of geographical information to close the loop of interactions between GIS and Societies.

1.8. Bibliography

[ANT 91] J.C. ANTENUCCI, K. BROWN, P.L. CROSWELL, M.J. KEVANY, H. ARCHER, *Geographic Information Systems: A Guide to the Technology*, New York, Van Nostrand Reinhold, 1991.

[ARM 68] S. ARMS, "Computer mapping in selected geographic information systems", *Proceedings Annual Meeting URISA,* Clayton, MO, 1968, 218-221.

[BAR 74] B. BARNES, *Scientific Knowledge and Sociological Theory,* London, Routledge & Kegan Paul, 1974.

[BEN 80] J.L. BENTLEY, "Multidimensional divide and conquer", *Communications of the Association of Computing Machinery,* vol. 23, 214-229, 1980.

[BER 73] B.J.L. BERRY, "Paradigms for modern geography", in R.J. CHORLEY (ed.), *Directions in Geography,* London, Methuen, 1973, 3-22.

[BIJ 87] W.E. BIJKER, T.P. HUGHES, T.J. FINCH (eds.), *The Social Construction of Technological Systems: New Directions in the Sociology and History of Technology,* Cambridge, MA, MIT Press, 1987.

[BIJ 92] W.E. BIJKER, J. LAW (eds.), *Shaping Technology/Building Society,* Cambridge, MA, MIT Press, 1992.

[BLO 76] D. BLOOR, *Knowledge and Social Imagery,* London, Routledge & Kegan Paul, 1976.

[BOW 95] G. BOWKER, "Manufacturing truth: The development of industrial research", in M. SERRES (ed.), *A History of Scientific Thought: Elements of a History of Science,* Oxford, Blackwell, 1995, 583-610.

[BUR 86] P.A. BURROUGH, *Principles of Geographical Information Systems for Land Resource Assessment,* Oxford, Clarendon Press, 1986.

[CAL 81] M. CALLON, "Pour une sociologie des controverses techniques", *Fundamenta Scientiae,* vol. 2, 381-399, 1981.

[CAM 93] I. CAMPARI, A.U. FRANK, "Cultural differences in GIS: A basic approach", *Proceedings of the 3rd European Conference on GIS*, Genoa, 1993.

[CAR 66] R. CARNAP, *Philosophical Foundations of Physics: An Introduction to the Philosophy of Science,* New York, Basic Books, 1966.

[CHR 75] N.R. CHRISMAN, "Topological data structures for geographic representation", *Proceedings AUTO–CARTO II,* 1975, 346-351.

[CHR 87] N.R. CHRISMAN, "Design of information systems based on social and cultural goals", *Photogrammetric Engineering and Remote Sensing,* vol. 53, 1367-1370, 1987.

[CHR 91] N.R. CHRISMAN, "Building a geography of cartography: Cartographic institutions in cultural context", *Proceedings International Cartographic Association 15th Conference,* Bournemouth, UK, 1991, 83-92.

[CHR 92a] N.R. CHRISMAN, "Ethics for the practitioners of Geographic Information Systems embedded in 'Real World' constraints of guilds, professions and institutional sponsorship", *Proceedings GIS/LIS 92,* 1992, 129-137.

[CHR 92] N.R. CHRISMAN, J.A. DOUGENIK, D. WHITE, "Lessons for the design of polygon overlay processing from the ODYSSEY WHIRLPOOL algorithm", *Proceedings 5th International Symposium on Spatial Data Handling,* 1992, 401-410.

[CHR 93] N.R. CHRISMAN, "Beyond spatio-temporal data models: A model of GIS as a technology embedded in historical context", *Proceedings Auto-Carto 11,* 1993, 23-32.

[CHR 96] N.R. CHRISMAN, "GIS as social practice", in T. HARRIS, D. WEINER (eds.), *Technical Report 96-7,* Santa Barbara, National Center for Geographic Information and Analysis, 1996, D9-D11.

[CHR 97] N.R. CHRISMAN, "Academic origins of GIS", in T. FORESMAN (ed.), *History of GIS,* London, Taylor & Francis, 1997, 33-43.

[CHR 99] N.R. CHRISMAN, "Speaking truth to power: An agenda for change", in K. Lowell (ed.), *Proceedings 3rd International Symposium on Accuracy of Spatial Databases Applied to Natural Resource Issues,* Ann Arbor, MI, Ann Arbor Press, 1999.

[CHR 01] N.R. CHRISMAN, "Configuring the user: social divisions of labor in GIS software", Presented at GIS in Society Workshop, Columbus, OH, 2001.

[COL 75] H.M. COLLINS, "The seven sexes: a study of the sociology of a phenomenon, or the replication of experiments in physics", *Sociology,* vol. 9, 205-224, 1975.

[COL 81] H.M. COLLINS, "Understanding science", *Fundamentals of Science,* vol. 2, 367-380, 1981.

[COW 88] D.J. COWEN, "GIS versus CAD versus DBMS: What are the differences?", *Photogrammetric Engineering and Remote Sensing,* vol. 54, 1551-1555, 1988.

[CRO 91] P. CROSWELL, "Obstacles to GIS implementation and guidelines to increase the opportunities for success", *Journal of the Urban and Regional Information Systems Association,* vol. 3, 43-56, 1991.

[CUR 91] M.R. CURRY, "On the possibility of ethics in geography: Writing citing and the construction of intellectual property", *Progress in Human Geography,* vol. 15, 125-147, 1991.

[CUR 95] M.R. CURRY, "GIS and the inevitability of ethical inconsistency", in J. Pickles (ed.), *Ground Truth: Social Implications of Geographic Information Systems,* New York, Guilford, 1995, 68-87.

[CUR 98] M.R. CURRY, *Digital Places: Living with Geographic Information Technologies,* New York, Routledge, 1998.

[DAN 84] J. DANGERMOND, C. FREEDMAN, "Findings regarding a conceptual model of a municipal data base and implications for software design", *Proceedings of the International Symposium on Spatial Data Handling,* Zurich, Switzerland, 1984, 479-496.

[DEL 76] G. DELEUZE, F. GUATTARI, *Rhizome Introduction,* Paris, Editions de Minuit, 1976.

[DEM 98] D. DEMERIT, "Science, social constructivism and nature", in B. Braun, N. CASTREE (eds.), *Remaking Reality: Nature at the Millenium,* London, Routledge, 1998, 173-193.

[DOB 83] J. DOBSON, "Automated geography", *Professional Geographer,* vol. 35, 135-143, 1983.

[DOB 93] J. DOBSON, "The geographic revolution: A retrospective on the age of automated geography", *Professional Geographer,* vol. 45, 431-439, 1993.

[DOU 80] J.A. DOUGENIK, "WHIRLPOOL: a geometric processor for polygon coverage data", *Proceedings AUTO-CARTO IV,* 1980, 304-311.

[FEE 95] A. FEENBERG, "Subversive rationalization: Technology, power, and democracy", in A. FEENBERG, A. HANNAY (eds.), *Technology and the Politics of Knowledge,* Bloomington, Indiana University Press, 1995, 3-22.

[FEY 93] P.A. FEYERABEND, *Against Method: Outline of an Anarchistic Theory of Knowledge,* London, Verso, 1993.

[FOR 97] T. FORESMAN (ed.), *History of GIS,* London, Taylor & Francis, 1997.

[FOU 77] M. FOUCAULT, *Discipline and Punish: The Birth of the Prison,* New York, Random House, 1977.

[FUJ 92] J.H. FUJIMURA, "Crafting science: Standardized packages, boundary objects, and "translation", in A. Pickering (ed.), *Science as Practice and Culture,* Chicago, University of Chicago Press, 1992, 168-211.

[GAL 97] P. GALISON, *Image and Logic,* Chicago, Chicago University Press, 1997.

[GOO 78] M.F. GOODCHILD, "Statistical aspects of the polygon overlay problem", in G. Dutton (ed.), *Harvard Papers on Geographic Information Systems,* Reading, MA, Addison Wesley, 1978.

[GOO 87] M.F. GOODCHILD, "A spatial analytical perspective on geographical information systems", *International Journal of Geographical Information Systems,* vol. 1, 327-334, 1987.

[GOO 92] M.F. GOODCHILD, "Geographical information science", *International Journal of Geographical Information Systems,* vol. 6, 31-45, 1992.

[GOO 95] M.F. GOODCHILD, "GIS and geographic research", in J. Pickles (ed.), *Ground Truth: The Social Implications of Geographic Information Systems,* New York, Guilford, 1995, 31-50.

[GOO 95a] M.F. GOODCHILD, D.M. MARK, M. EGENHOFER, E. SHEPPARD, *Advancing Geographic Information Science: A Proposal to National Science Foundation for Project Varenius,* Santa Barbara, University of California, 1995.

[GOS 95] J. GOSS, "Marketing the new marketing: The strategic discourse of geodemographic information systems", in J. Pickles (ed.), *Ground Truth: Social Implications of Geographic Information Systems,* New York, Guilford, 1995, 130-170.

[HAR 89] J.B. HARLEY, "Deconstructing the map", *Cartographica,* vol. 26, 1-20, 1989.

[HAR 90] J.B. HARLEY, "Cartography, ethics and social theory", *Cartographica*, vol. 27, 1-23, 1990.

[HAR 96] T. HARRIS, D. WEINER (eds.), *Scientific Report for the Initiative 19 Specialist Meeting,* South Haven, Minnesota, Santa Barbara, National Center for Geographic Information and Analysis, March 2-5, 1996.

[HAR 98] F. HARVEY, N.R. CHRISMAN, "Boundary objects and the social construction of GIS technology", *Environment and Planning A,* vol. 30, 1683-1694, 1998.

[HAR 00] F. HARVEY, "The social construction of geographical information systems", *International Journal of Geographical Information Science,* vol. 14, 711-713, 2000.

[HER 87] J. HERRING, "TIGRIS: Topologically integrated geographic information system", *Proceedings AUTO-CARTO 8,* 1987, 282-291.

[HOP 77] L.D. HOPKINS, "Methods of generating land suitability maps: A comparative evaluation", *American Institute of Planners Journal*, vol. 43, 386-400, 1977.

[HUT 95] E. HUTCHINS, *Cognition in the Wild,* Cambridge, MA, MIT Press, 1995.

[KNO 83] K. KNORR-CETINA, "Towards a constructivist interpretation of science", in K. KNORR-CETINA, M. MULKAY (eds.), *Science Observed: Perspectives on the Social Study of Science,* Beverly Hills, Sage, 1983.

[KUH 70] T.S. KUHN, *The Structure of Scientific Revolutions,* Chicago, University of Chicago Press, 1970.

[KWA 02] M.P. KWAN, "Feminist visualization: Reenvisioning GIS as a method in feminist geographic research", *Annals of the Association of American Geographers*, vol. 92, No. 3, 381-391, 2002.

[LAP 97] J. LAPLACA, "Nature park site analysis in Fairfax county Virginia using the arcview spatial analyst extension", *ESRI Users Conference,* Palm Springs, CA, 1997, Available: http://gis.esri.com/library/userconf/proc97/proc97/to550/pap517/p517.htm

[LAT 86] B. LATOUR, S. WOOLGAR, *Laboratory Life: The Construction of Scientific Facts,* Princeton, NJ, Princeton University Press, 1986.

[LAT 87] B. LATOUR, *Science in Action,* Cambridge, MA, Harvard University Press, 1987.

[LAT 90] B. LATOUR, "Drawing things together", in Lynch, Woolgar (eds.), *Representation in Scientific Practice,* Cambridge, MA, MIT Press, 1990, 19-68.

[LAT 93] B. LATOUR, *We Never Were Modern,* Cambridge, MA, Harvard University Press, 1993.

[LAT 96] B. LATOUR, *Aramis or the Love of Technology,* Cambridge, MA, Harvard University Press, 1996.

[LAT 99] B. LATOUR, *Pandora's Hope,* Cambridge, MA, Harvard University Press, 1999.

[MAC 90] P. MACK, *Viewing the Earth: The Social Construction of the Landsat Satellite System,* Cambridge, MA, MIT Press, 1990.

[MAK 90] D.A. MACKENZIE, Inventing accuracy: An historical sociology of nuclear missile guidance, Cambridge, MA, MIT Press, 1990.

[MAG 01] D. MAGUIRE, "GIS software: Progress and prospects", *Proceedings GISRUK,* Cardiff, 2001.

[MAR 00] E.W. MARTIN, "Actor-networks and implementation: Examples from conservation GIS in Ecuador", *International Journal of Geographical Information Science,* vol. 14, 715-738, 2000.

[MON 95] D.R. MONTELLO, S.M. FREUNDSCHUH, "Sources of spatial knowledge and their implications for GIS: an introduction", *Geographical Systems,* vol. 2, 169-176, 1995.

[MOR 94] J.L. MORRISON, "Cartography and organization", *ACSM Bulletin,* vol. 148, 38-45, 1994.

[NRC 83] NATIONAL RESEARCH COUNCIL, *Procedures and Standards for a Multipurpose Cadastre,* Washington, DC, National Academy Press, 1983.

[NIE 87] B.J. NIEMANN, JR., J.G. SULLIVAN, S.J. VENTURA, N.R. CHRISMAN, A.P. VONDEROHE, D.F. MEZERA, D.D. MOYER, "Results of the Dane County Land Records Project", *Photogrammetric Engineering and Remote Sensing*, vol. 53, 1371-1378, 1987.

[OBE 98] N. OBERMEYER (ed.), "Special issue on public participation GIS", *Cartography and Geographic Information Science,* vol. 25, 65-122, 1998.

[ONS 91] H.J. ONSRUD, J.K. PINTO, "Diffusion of geographic information innovations", *International Journal of Geographical Information Systems,* vol. 5, 447-467, 1991.

[OPE 91] S. OPENSHAW, "A view on the crisis in geography, or using GIS to put Humpty-Dumpty back together again", *Environment and Planning A,* vol. 23, 621-628, 1991.

[PEU 75] T.K. PEUCKER, N.R. CHRISMAN, "Cartographic data structures", *The American Cartographer,* vol. 2, 55-69, 1975.

[PIC 95] J. PICKLES (ed.), *Ground Truth: Social Implications of Geographic Information Systems,* New York, Guilford, 1995.

[PIC 97] J. PICKLES, "Tool or science? GIS, technoscience, and the theoretical turn", *Annals of the Association of American Geographers,* vol. 87, 363-372, 1997.

[RIA 95] A. RIAHI-BELKAOUI, *The Cultural Shaping of Accounting,* Westport, CT, Quorum Books, 1995.

[ROB 95] S.M. ROBERTS, R.H. SCHEIN, "Earth shattering: Global imagery and GIS", in J. Pickles (ed.), *Ground Truth: Social Implications of Geographic Information Systems,* New York, Guilford, 1995, 171-195.

[SCH 00] N. SCHUURMAN, "Trouble in the heartland: GIS and its critics in the 1990s", *Progress in Human Geography,* vol. 24, 4 (2000) pp. 569-590, 2002.

[SCH 02] N. SCHUURMAN, "Reconciling social constructivism and realism in GIS", *ACME,* ACME: An International E-Journal for Critical Geographies vol. 1(1): 75-90, 2002.

[SCO 98] J. SCOTT, *Seeing Like a State: How Certain Schemes to Improve the Human Condition Have Failed*, New Haven, Yale University Press, 1998.

[SHE 95] E. SHEPPARD, "GIS and society: Towards a research agenda", *Cartography and Geographic Information Systems,* vol. 22, 5-16, 1995.

[SMI 92] N. SMITH, "History and philosophy of geography: real wars, theory wars*",* *Progress in Human Geography*, vol. 16, 257-271, 1992.

[SNO 59] C.P. SNOW, *The Two Cultures and the Scientific Revolution,* Cambridge, Cambridge University Press, 1959.

[STA 89] S.L. STAR, J.R. GREISEMER, "Institutional ecology, "translations," and boundary objects: amateurs and professionals in Berkeley's museum of vertebrate zoology", *Social Studies of Science,* vol. 19, 387-420, 1989.

[STE 76] C. STEINITZ, P. PARKER, L. JORDAN, "Hand-drawn overlays: Their history and prospective uses", *Landscape Architecture,* vol. 66, 444-455, 1976.

[TAY 91] P. TAYLOR, M. OVERTON, "Commentary: Further thoughts on geography and GIS", *Environment and Planning A,* vol. 23, 1087-1094, 1991.

[TOM 90] C.D. TOMLIN, *Geographic Information Systems and Cartographic Modeling*, Englewood Cliffs, NJ, Prentice Hall, 1990.

[TOM 68] R.F. TOMLINSON, "A geographic information system for regional planning", in G.A. Stewart (ed.), *Land Evaluation,* Melbourne, Macmillan, 1968.

[TOM 84] R.F. TOMLINSON, "Geographic information systems – A new frontier", *Proceedings of the International Symposium on Spatial Data Handling,* Zurich, Switzerland, 1984, 1-14.

[TOM 89] R.F. TOMLINSON, "Presidential address: Geographic information systems and geographers in the 1990s", *Canadian Geographer,* vol. 33, 290-298, 1989.

[TOM 98] R.F. TOMLINSON, "The Canada geographic information system", in T.W. FORESMAN (ed.), *The History of Geographic Information Systems: Perspectives from the Pioneers,* Upper Saddle River, NJ, Prentice-Hall, 1998, 21-32.

[UCG 96] UCGIS, "Research priorities for geographic information science", *Cartography and Geograpic Information Systems,* vol. 23(3), 115-127, 1996.

[UCG 98] UCGIS, *Cognition of Geographic Information,* Available: http://ucgis.org/research_white/cog.html, 1998

[USH 54] A.P. USHER, *A History of Mechanical Inventions,* Cambridge, MA, Harvard Press, 1954.

[WAR 57] W. WARNTZ, "Transportation, social physics and the law of refraction", *Professional Geographer,* vol. 9, 2-7, 1957.

[WAR 65] W. WARNTZ, *A Note on Surfaces and Paths: Applications to Geographical Problems*, Michigan Inter-University Community of Mathematical Geographers, 1965.

[WAR 66] W. WARNTZ, "The topology of socio-economic terrain and spatial flows", *Papers, Regional Science Association,* vol. 17, 47-61, 1966.

[WOO 92] D. WOOD, *Power of Maps,* New York, Guilford, 1992.

[WOO 87] S. WOOLGAR, "Reconstructing man and machine: A note on sociological critiques of cognitivism", W.E. BIJKER, T.P. HUGHES, T.J. FINCH (eds.), *The Social Construction of Technological Systems: New Directions in the Sociology and History of Technology,* Cambridge, MA, MIT Press, 1987, 311-328.

Chapter 2

The Spacc and Place Trades versus Geographic Information Technologies: Babel and Esperanto

2.1. Introduction

Geographic information science (GIS) is a convenient catch-all term. Its practitioners believe that it has become, in technical terms, a body of perfectly integrated methods and tools and, in terms of use, a homogenous community of experts. In practice, though, GIS, or more globally, geographic information technologies (GIT), is practiced in a plurality of forms for at least three "organic" reasons:

– geographic information (GI) itself (whether in digital form or not) is closely related to the public object, and consequently, to the political institutions and technical bodies that frame and use it;

– through scale and quality issues, GI refers to mental and physical models that are theoretically nested (packed together closely) or contiguous, but actually largely distinct and impenetrable;

– GI provides uses from the most detailed modeling applications to the widest synthesis of technologies and concepts.

Chapter written by Grégoire FEYT.

The combination of these aspects is what makes GI a problematic *notion* and a particularly difficult object of inquiry. This "organic" complexity appears through operational plans and through a mosaic of trades, producing or using geographical data in contexts and in extraordinarily diversified perspectives "since the Earth was territory".

2.1.1. *The "historical mission" of GISs*

During the last 15 years, "GIS business" has gradually emerged within this mosaic. However, a view of the general and historical process of integration of computer tools into professional practices and the problems of territory business with GIS remain persistently current. This would not in itself be particularly disconcerting if we had not, right from the start, reversed the proposition (GIS facing the territory trades) by entrusting to GIT the mission of communication among disciplines and trades, which until now attempted to make a coexistence between their disciplinary languages and a modeling and representation of the "real world".

However, after nearly 20 years, we must admit that, faced with the Babel of the space and place trades, it is difficult to produce the desired effects with the use of "technical and methodological Esperanto". We can certainly choose to blame this delay on insufficient scientific and technological development. Conversely, we can explain this by the fact that GIS has somewhat been a victim of its own success. Most businesses using georeferenced data have quickly been able to find, in GIS formalism and functions, the resources needed to automate and optimize all or a part of their expertise and know-how. This is true except that, through authorizing forms of communication impossible until now, the adoption of a common language can produce paradoxical effects. Whereas we cannot, literally, listen to or understand each other, we can still start with the principle that, fundamentally, "we understand each other" and "we listen to each other". By using a single language or, in this case, a single formalism, we can realize, progressively or brutally, that words and models can contain meanings and values that are different, if not opposite.

2.1.2. *GIS, first a "machine raising questions"*

The hypothesis adopted here along with others [MAJ 99] is that, by wanting to be integrated, GITs have actually contributed to highlighting, revealing or exaggerating differences in the way that territory businesses understand and model space by referring to a system of values and to a specific business-culture. This is undoubtedly a greatly unpredictable problem that has progressively grown in importance during the expansion of the field of application and appropriation of GISs by the many businesses involved. Still, this particular test in the course of GIT generalization does not put into question the "federating mission" that was historically and organically assigned to them, and in fact it is quite the opposite. In fact, by "touching the raw nerve" in the vital process of spatial and territorial expertise sharing, GITs play a new role, an especially efficient role, insofar as these technologies can often bring operational answers to theoretical or organizational questions that they helped to raise.

To do this, GIS engineering must actually consider the situation of culture-business where the substance is generally based on historical and cultural factors and scientific and technical considerations, as highlighted in Chapter 1. We must also admit once and for all that GI is not quite common information and that the process of digitization and dematerialization cannot, as was the case with the other forms of information (numbers, text, image, and sound), be fixed in the framework of the social or professional demand/technological progress-competitive economy dialectic.

In this perspective, this second chapter is intended as an overview of the relationships between territory trades with, on one side the historical, technical, and political contexts in which they fit, and, on the other side, technical and technological evolutions to which they were, and still are, compared for the last two or three decades. The approach proposed is somewhat cartographic insofar as it tries to identify in a general way first, then through consecutive "zoom strikes", the facets that seemed the most significant in the context of this book.

2.2. Interface trades

2.2.1. *A complex system*

The problem of GIS facing territory trades, the reverse is also a problem, brings into question at least four distinct and relatively non-correlated aspects in terms of historical evolution as well as operational methods and value systems.

2.2.1.1. *Territory trades*

This very generic terminology covers the fields of competence, fields of application, and very diversified practices. From the topographer creating an urban database, to the ecologist developing environmental indicators, to the urban planner triggering a debate around an urban planning local plan (UPLP), to the consulting firm attempting to optimize public transit, the common point, and this is significant, resides in managing or enhancing the asset and common construction resources, known as territory, as best as possible. There are many complex differences among these practices, however: training, formalism, levels of hierarchy, geographic environment, nature of themes, institutional framework, etc. Within this professional patchwork, which is only a reflection of the effects of a territorial complexity, GITs occupy a discrete but vague position. The question is to determine how this potentially strategic scattering can enable GITs to play a specific role in the creation of an operational connection between "business-cultures", where all professional and territorial stakeholders recognize and typically bemoan segmentation without really knowing how to stop it.

2.2.1.2. *Representation of space*

By using the term geographic information, the geomatic community means to include all the spatial data that can possibly be used in a digital form. In so doing, the community tends to forget the fact that, in the minds of non-specialists, the preferred form of expression of spatial phenomena remains the map. It simultaneously benefits and suffers in public opinion and in general practice in its status as a universal language that is familiar and "objective", a status that paradoxically controls the growth of GITs in terms of methodological innovation. This fact is highly significant. First, the roots of maps are deep and go back to basic learning and our collective history. Second, most sponsors, as well as the public that professionals who use geographic data deal with (political, administrative or economic decision-makers, socioeconomic stakeholders, residents, etc.), are, and will be for a long time, "lay people" for whom the map, in its most traditional form, constitutes the alpha and omega of spatial expertise.

2.2.1.3. *Information technologies*

Globally, the development and "popularization" of information technology, in the broad sense of the term, were mainly guided by a combination of commercial competition and technological progress, or, in other words, according to mostly financial criteria. We must admit that, in this context, the GI sector has always been an "economic dwarf" compared with the major fields of digital technology (text, sound, image), but a clever dwarf nonetheless, who has constantly been able to take advantage of technological, or even methodological, innovations created and financed by other sectors and for other audiences. Here again, the penetration of GITs within businesses is obvious. The developmental history of GISs has widely been described by handymen and enthusiastic pioneers (see Chapter 1), and we can imagine that at least one aspect of future innovations, – whether are methods or uses, will continue to operate the same way.

2.2.1.4. *Management and public decision*

Even though, at least until very recently, GITs were not considered a "mass public" technology, they were still, and greatly remain, by their nature and in their culture, a technology at the service of the common good. In fact, the main production and use of geographic data is managed by and for public or para-governmental organizations (government decentralized services theory nesting or contiguous but largely distinct and impenetrable, territorial communities, public or para-governmental establishments, consular divisions, etc). GITs must face the problem of organizational operations as well as the institutional and political operation of territories. In terms of technical implementation, this is expressed through the problem in defining quality-objectives for organizational GISs. By formalizing, and by freezing, the description of a "territorial reality" through a data model, by revealing spatial structures or unsuspected or undisclosed trends, or by introducing precision where we have collectively settled for vagueness, GISs often have to play a new and sometimes disturbing role in the political process. Applied to many businesses, this observation returns to an additional condition assigned to the development of GITs, which was to invent methods and practices of figuring, explaining, and adapting, making the appropriation of territorial expertise by stakeholders and decision makers possible.

2.2.2. *A system under pressure*

Positioned by nature at the juncture between technical and technological, political and cultural problems, territory trades are inevitably stuck in the movement of territories. This book is not intended to go into great detail on the processes and methods of reconstructing the territorial organization implemented in the French territories. We will only cover in this chapter the aspects directly involving professional practices in their relation to tools and methods of a knowledge of the territory.

2.2.2.1. *The multiplication and specialization of territory trades*

For a few decades now, we have been witnessing, in the territorial field as in other socioeconomic sectors, a multiplication and increasing specialization of trades and professional managers. The reasons are multiple, and, here again, we will only mention a few elements of this discussion that will necessarily be incomplete.

The main reason for this multiplication is the obvious scientific and technical progress which, by multiplying and improving on their fields of expertise, have triggered the consolidation of professional sectors within their own formalism, nomenclature, data, tools, etc.

Another explanation, linked to the previous one, goes back to the complexity and technical vulnerability of "modern" society. The importance, in all senses of the term, of infrastructures that we depend on (transportation, land improvement, consumption, product supply, water, electricity, information, etc.) requires the use of specialized professionals using methods, tools, languages, and data that are increasingly more detailed and secure.

This functional complexity also comes with a significant development of technical and legal standards and contexts. Professionally, this evolution is expressed by the appearance of field of expertise connected to functions of creation as well as technical and regulatory control. In terms of description and representation of objects and the spatial processes involved, we can easily imagine that prior modeling and subsequent evaluation required data, methods, and, in a more general way, technical postures of different natures.

In short, we suggest the idea that, if we use the image of a tree, spatial expertise and its expression on different disciplinary and informational levels divide into increasingly numerous, deep, and finite root, roots that grow farther and farther apart from each other.

2.2.2.2. *The map, a dominating but unsatisfactory mode of comprehension of space and territory*

If, to continue the tree metaphor, we "return to the surface" and question "in the light of day" the varying branches of expertise, we can observe that the territory trades are stuck with changes that pertain to the synthetic and no longer, as was the case in the previous section, to the analytical.

As previously indicated, the cartographic representation in its historical form remains the dominant mode of representation of space and its dynamics. In fact, with a few exceptions, the structures of geographic data and the most elaborate of spatial analysis methods show *in fine* through cartographic conventions and methods inherited from the plan and the paper map. The reason for this "conservatism" is not to be found scientifically and technically; it goes back to the fact that the groups using geographic data are, in terms of decision or control, those (in the best of cases) accustomed to the traditional cartographic language.

But this language seems in crisis as well [DEB 02]. In fact, the "traditional" mode of representation, that of static figuration, of the 2D orthogonal mark and the Euclidian metric, is increasingly unable to account for the nature and operation of infrastructures previously mentioned or their collective and individual spatial practices [LEV 02]. What has long been an area of dissatisfaction, flaws formulated by and for specialists, now seems to be one that is increasingly found difficult to cope with by the public.

Increasingly asked to communicate their territorial expertise, the territory trades are in a sense "betrayed" by the methods of representation that we want them to use. The connection with GITs is obvious here on two levels. First, this is because of the "cognitive prejudice" resulting from the shift from a formalism based on trade data with its generic and universal language that the map represents. Also, however, paired with information and communication technologies, GITs open up perspectives to a promising obviousness that will clearly disrupt the mode of relation historically established between the expert, the sponsor, and territorial stakeholders around the understanding and use of the cartographic representation [DEB 03]. From this point of view, it is obvious that recent development and the increasing popularization of Webcarto and virtual navigation tools are bringing out a new generation of uses and stakeholders in territorial knowledge.

Territory trades, who used to be the only legitimate producers of cartographic information, are now facing another interpellation: stakeholders in the civilian society are now able to access resources (hardware, software, and data) that make it possible for them to produce their own maps, virtual composites, and other geowikis that can compete as an expression of spatial (counter-)expertise.

More generally, in the question of (carto)graphic representation, the issue of the status of expertise and the *expert* is raised within the territory trades. For a long time, the plan and map were designed and perceived either as strictly technical or scientific documents, or as an almost symbolic figuration irrefutable in any case of knowledge or power. Rightly or wrongly, this form of expression of spatial expertise is now questioned more frequently by territorial stakeholders; because of this, it is the expertise itself that is attacked, insofar as it is located or "brought" to a land where its specialized formalism or language is quickly brought into question. Territory trades, not able, or even wanting, to "desert" this field, are thus required to rethink the way they communicate (about) their expertise based on modes of expression and confrontation that can obviously no longer be unidirectional or univocal.

2.2.2.3. *Generalization of information technology and the proliferation of databases*

Originally, the business of a small circle of all-powerful insiders managing complex and expensive systems, including the professional sphere, information technology has gradually become individualized and distributed to the point where each practitioner, or even citizen, has a computer. Management and operations software, alphanumeric in the beginning and now also cartographic, followed the same quantitative and qualitative evolution.

We have witnessed in the last 10 years a multitude of territorial databases, and thus knowledge, created by following a logic of different objectives: from the practitioner wanting to "put some order" in the accumulation of acquired experience and knowledge into a given sector or territory, to the observatory created because of public policy, or better still the department concerned about mutualizing and capitalizing on information required for its missions. These initiatives, most often carried out with intelligence and determination, but without coordination or structure, have led to a true "geomatic Balkanization" that we are only now starting to measure, or even feel, the effects of. How many territorial institutions are even able to list all the databases they host?

Without dwelling on the sometimes "private" character of a database created by or for an agency or department (and often disappearing with it), the problem in terms of global territorial expertise arises from the fact that each of these applications was undoubtedly created according to a specific *vision* and *model* that are, in a word, independent. Because of this, the informatization process only reproduced the preexisting scientific, professional, and institutional divides brought up above. However, once "budgeted" for, modeled, implemented, installed, and "appropriated" by its user(s), a database often tends to be perceived as "cast in stone," inducing rigidity and inertia where we were expecting flexibility and reactivity. Scientific and technological developments (generic data models, interoperability, cartographic generalization, etc.) obviously will only be able to partially remedy this compartmentalization insofar as it starts with differences in the mental models in territorial disciplines.

At a time where territory trades have to face transversality and increasing legibility requirements, this legacy of the recent past is starting to weigh heavy on them. The development of information technologies enables the reduction of business-expertise requirements but, at the same time, tends to penalize the debates around this same expertise. To use the image of roots again, it is as if the growth of computer tools was like fertilizer, making it possible for the root system to grow deeper and faster into the field of spatial and territorial knowledge. Most professionals are now conscious of the scope of the problem and its challenges. The studies and initiatives being performed, notably in terms of cartographic reference[1] or "technico-institutional" devices of resource mutualization[2] are drawing up situations to which the territory trades will have to commit, with more or less ease and spontaneity.

---

1 Such as the Référentiel à Grande Echelle (RGE Large Scale Reference) project for proposing a coherent base established from topographic elements deemed of general use for large and medium scale applications for the French territory.

2 For example, in the image of the multi-partner resource center created by the Provence-Alpes-Côte d'Azur region to enable public and private stakeholders to access reference geographic data (CRIGE-PACA).

2.2.2.4. *Trades in the process of territorial mutation and recomposition*

Territory trades can hardly escape the pressures, impulses, and mutations concerned with this *living organism* that is territory, one complex in its "physiology" and in its "psychology". In France, this organization is actually disrupted by interrogations and disturbances relative to its mode of organization and operation. The administrative and technical context inherited from the French Revolution, that of a centralized government operating its direct or indirect authority on territorial levels or fields that are identified, prioritized, and stable, has for the last 30 years been in the grip of resistance, and the outcome is still uncertain.

Decentralization, past, current and future, the European reconstruction, the place and increasing strategic role of agglomeration, competition between territories, all of these gradually or brutally disrupt the framework of control both formal and implicit to which the major part of the territory trades are subject. The withdrawal of the government as a technical operator, the careful creation of a technical expertise and legitimacy between local communities, the approximate cartography of the distribution of skill sets between territorial institutions (government, different levels of territorial community, public institutions of cooperation between local communities) have a paradoxical effect in terms of the territory trades. While making traditional tasks more crucial and tricky, these multiple evolutions enable the emergence of new missions linked to the appearance or revelation of "dead zones" or "terrae incognitae" in terms of territorial knowledge and expertise.

Through the few points summarily discussed here, we can see how the territory trades are, today, in the middle of a "problem mix" combining constraints from "traditional" technologies to interrogations and challenges linked to an increasing requirement of transversality and public debate.

To a large part, the answers, or even the initiative of the response itself, to the many questions raised will first involve the contracting authority and usage authority, or, in other words, the political and public spheres. On this subject, a true awareness of the strategic dimension, and one no longer strictly technical, of GI and its technologies are beginning to emerge among French public decision makers at the governmental level [JAI 04] as well as in the territorial communities. The implementation of cartographic portals, for example, by the National Geographic Institute or by regional councils, expresses and encourages a "virtuous" evolution toward the implementation of information-sharing devices. Because of this, the increasing popularity and performance of commercial or unrestrained portals, which are increasingly involved in the territorial debate, or even controversy, do not offer institutional stakeholders much choice but to "follow the movement," given that they could not lead it. In this global site, territory trades, including GIS, play their part, particularly in the strategic aspects, where technicality or hermetic quality does not lay itself open to interpellation or public initiative (interdisciplinarity, interoperability, semiology, spatial analysis, collaborative uses, etc.).

In this field, however, scientific progress was always greatly linked to the diversification of uses, which in turn is linked to the development of technologies. It then seems difficult to only consider the emergence and growth of GITs as a new and exogenous resource, of "deus ex machina", that territory trades would only have to control or adapt. The elements and history of the problem are also complex and require that we spend more time explaining them to better understand.

2.3. The long progress toward the culture of the digital GI

The history of GITs was not written on a blank page (see Chapter 1). As with territory trades that exist in a context that greatly overwhelms them, the development of GITs fits into a movement that incorporates scientific and technical progress at the same time that it is part of a competitive economy. However, as most tertiary and engineering trades seem to have achieved and integrated this technological mutation quickly and easily, GI and its trades have a hard time, or at least have taken a long time, entering the digital age. The lack of truly functional transversal GISs is obvious at least in terms of the expressed or latent needs and perspectives announced a long time ago. To understand and overcome these problems, it is important to go back to the process of GIT penetration and distribution within a professional world that is rich and "heavy" in its diversity and institutional and technical history.

2.3.1. *GIT and trades: a culture of pioneers*

2.3.1.1. *Ancient history*

2.3.1.1.1. In the beginning, there was CAD

As Mister Jourdain was making prose without realizing it, the first users of what was not immediately called "*GIS*" were manipulating geographic data while believing they were drawing. However, in an era in which computers were expensive, resource intensive, and required rooms and experts dedicated to their use, only large organizations could afford the "luxury" of migrating their plan or map production from the drawing board to the screen (alphanumeric and black and white in these proto-historic times). Except for the military, the technical departments of large cities and large network operators, for whom the financial burden created by digitization was relatively light compared to the cost and global stakes of their activity, were mainly the ones in France that embarked on the course of what was originally thought to be, and designed as, computer-assisted design (CAD). "The system creating the function", the gradual familiarity with the tool, stimulated by the progress of computers and software, triggered the development of increasingly elaborate functions evolving toward concepts of urban databases, followed by GIS.

This "technological" origin profoundly marked the development of GITs, not only in terms of its fields of use, but also in the ergonomics and design of software and data models. Thus, by only considering the vector model and by favoring quality criteria mainly involving topology or input constraints, software editors, thus responding to the expectations of their main clients of the moment, have somewhat curbed the distribution of the use of GISs out of their "historical biotope". Some potential fields of application, notably the environmental field, have had no other alternative but to "abstain" or compromise with modes of analysis that are not well adapted to the nature of their problems or to the characteristics of their business data.

This situation started to emerge about six years ago because of the integration of methods linked, for example, to spatial analysis or image processing in trade or generic GISs, and is now used in research or through specialized or "confidential" applications. Nonetheless, this historical determinism has weighed and continues to weigh on the capacity of GITs to structure a transversal approach in the public sphere as well as in scientific research.

2.3.1.1.2. A GI complexity commonly underestimated

This culture of design and management has greatly influenced from the beginning our understanding of what the term GI means, perceived even then as digital.

The "historical" and still greatly dominant fields of application of GISs concern objects or functions involving project management, or, in other words, problems that we can qualify or measure the ins and outs of. The modeling of a land improvement network or the creation of cadastral database requires that we comply as closely as possible with the technical criteria that governed the design or definition of spatial objects and their respective properties. Even if the implementation of the methodology can create formidable problems, the purpose and objectives of quality are such that they can be globally identified. The problems are revealed to be more complex when we have to model a physical or social environment in the construction and operation from which notions of scale, or topology and typology, do not have (pre)defined existence or relevance.

The dematerialization of cartographic documents, operating out of concepts and methods from the CAD world and alphanumeric databases, quickly revealed characteristics or "gray zones" of GI that cartographic design, in its traditional or even artistic dimension, had until then made it possible to process, avoid or overshadow [ECO 94]. The implicit understanding that the cartographer's brain/hand relationship gives to the reader's eye/brain relationship disappears with the implementation of a formal model, even as elaborate as this can be. What is the relevant scale to model an environmental variable? Does a row of plane trees along a highway define a linear or a series of occurrences? How far should we push the sharing of topology between layers? What is a true land register? These questions have only become relevant immediately before (or after) digitization.

Questions emerging from or because of the materialization of cartographic data again challenged the territory trades in their practices and representations of space. Beyond the strictly technical or financial aspects, the very concrete problems identified over time by the producers and users of digital geographic data have followed, sometimes very crudely, the differences in terms of hierarchy, quality, or even ethics in the way of understanding and considering the territory. Paradoxically, it may be this involuntary capacity of GISs to reveal differences that makes them precious tools and levers in the search for a comparison between themes and professional practices. By making abstract or incantatory references obsolete to transversality, GITs force the identification and formalization of what makes sense in common usage from a constant dialectic between modeling, production, and data use.

2.3.1.1.3. No GIS without reference data

This search for shared geographic references did not appear at the birth of GISs. First used as *business* tools, businesses were not satisfied for long with an ability to show color in the polylines or polygons as their only skill sets. To suit the documents produced and to have a background on which the different themes can position themselves, the need for "general interest" geographic data has appeared.

As in so many other fields, this notion of general interest goes back to the question of the status of legitimate operator(s) to provide what resembles a *service* and no longer "simple" expertise. This question is still not completely settled in France, even though we thought it had been several times. Long controlled by the large public organizations and the professional stakeholders organically concerned, the available reference geographic data has considerably evolved, in quantitative and qualitative terms, obviously because of the emergence of private operators who are often international but especially because of the pressure of users or clients who are increasingly experienced and well equipped.

That is certainly the question of references that have more clearly occurred within the GIS trades, between those working *for* GISs and those working *with* GISs. In fact, the notion of background, richer and more consensual than cartographic reference, goes back to registries exceeding the strict framework of professional practices, including in their comparison. This background is first used to understand and to know. Because of this, it is linked to notions of territorial identity and practice, where assessment criteria do not fit very well with those at the foundation of GIS performance.

Because of, and in the name of, these different points of view, a number of misunderstandings and dissatisfactions between practitioners formulating needs considered trivial or minor by GIS specialists have arisen, emphasizing the specifications perceived as too "precise" by these same practitioners. Beyond its anecdotal aspect, this question of geographic reference(s) goes back once again to the question already mentioned above of the criteria of quality according to and from which the territory trades understand space and the data that represent it.

2.3.1.2. *An arduous history*

The term GIS is only now starting to become popular among an audience who knows about, and who has even been using for a long time, digital sound and image technologies, although these emerged after GI technologies. A major paradox undoubtedly resides in the fact that, finally, the public will have access to GI without even having to know anything about GIS because of the Internet and GPS systems. In fact, GISs until now have remained the business of institutions, professionals, and enthusiasts, and we can wonder in what measure they could not be ontologically destined to remain that way.

2.3.1.2.1. GIS, long the business of institutions, specialists, and enthusiasts

Scientific advances and methods are greatly built on the fly, based on the diversification of uses and needs. What's more, computer technologies are constantly evolving in the heterogenity of professional and institutional stakeholders. All these ingredients have been in place for a long time and restrict initiative and control of GIS development within businesses and organizations with a direct or central interest. However, this does not include those within the territory trades who had the "revelation" of GISs to the point of acquiring, mainly by their own initiative, the control of "gas plants" or "traditional machines" available until the end of the 1980s.

With the advent of technological progress and the laws of the free market, the offer of software has today been "clarified". Nonetheless, this culture of "pioneers" and "handymen" deeply marked and continues to mark the development of GISs. This seems inevitable insofar as it reflects the evolution of businesses and its problems involved.

2.3.1.2.2. For the longest time, "on-the-job" training for GITs

The major problem that the development and distribution of GITs faced is training. Two categories in this problem must be defined: specialized training and general training, what we might call "geomatic education".

In terms of the first category, the development occurred during the last decade, and there is no longer much training in businesses using geographic data that does not include at least an initiation to GISs or automatic mapping. The question of "general culture" for GI and its technologies is more tricky, however. Whereas the map is a support for information involving, directly or indirectly, a large quantity of disciplines and activities, GISs remain, in major part, either unknown to most or perceived as a tool for a small group of "trendy" practitioners or specialized technical people. This is particularly relevant in the fields of management and restricts accurate development, or even the development at all, of organizational GISs.

2.4. Temporary conclusion

The territory trades appear to be subjected to a series of constraints and explicit or latent requests greatly exceeding the context of their strict professional and technical nature. A common denominator nevertheless exists: the question of territorial expertise, sharing, comparison, and transmission. Beyond this, the notion of information is involved and the responses to it necessarily change, with improvements in design, treatment, and representation of this notion. The new capabilities of GITs have actually reached a point at which they can confront the increasing complexity of the operation, management, and governing of territories.

At the same time, these methods and technologies, because they concretely raise questions that had remained formal or hypothetical until now, have a tendency to make each field of expertise return to its own representations of space. Business applications have thus continued their development such that the implementation of transversal or organizational GISs must deal with problems that it should not have to deal with.

Because of this difficult and unrewarding situation for GIS experts or promoters, they might give in to the temptation of a type of geomatic "self-sufficiency" or "assertion" that could lead to a segmentation of territorial knowledge that would be more counter-productive than occurred during the pre-computer era. To respond to their "historical mission" of being integrating tools, GITs and their users (specialists and practitioners) must achieve a new reasoning in terms of problematics and not simply problems, and, on the other hand integrating, without necessarily adopting, the methods and types of representation of the territorial stakeholders.

2.5. Bibliography

[DEB 02] B. DEBARBIEUX, M. VANIER (ed.), *Ces Territorialités qui se Dessinent,* La Tour d'Aigues, Editions de l'Aube, 2002

[DEB 03] B. DEBARBIEUX, S. LARDON (ed.), *Les Figures du Projet Territorial,* Editions de l'Aube/Datar, 2003

[ECO 94] C. ECOBICHON, *L'information Géographique. Nouvelles Techniques, Nouvelles Pratiques*, Hermès, 1994

[JAI 04] M.Ch. JAILLET *et al.*, "L'évaluation de la connaissance du territoire par les services du ministère de l'Equipement," report written from a request of the Ministère de l'Equipement by the Conseil Général des Ponts et Chaussées, 2004, available at: www2.equipement.gouv.fr/rapports/archive_r/adm_r.htm

[LEV 02] J. LÉVY (collective work under the direction of), Jeu de cartes, nouvelle donne: cartographier aujourd'hui les espaces d'aujourd'hui, research report from the Cartogram group for DATAR, Tours, 2002

[MAJ 99] W. MAJOR, *Approche de la concertation territoriale par l'analyse systémique et l'analyse lexicale du discours des acteurs. Perspectives d'application aux systèmes d'information géographique*, Ecole Polytechnique Fédérale de Lausanne, 1999

Chapter 3

The Future of GIS from an IT Evolution Perspective

3.1. Introduction

A quick analysis of the different trends in information technologies (ITs) brings out the elements that should be developed in the field of geographical information systems (GISs) to have better integration of their elements with current applications and during the development of organizational administrative systems.

In most IT applications in companies and public organizations, there is a need to integrate data from several sources; this need has often been the trigger for new applications. Because of high-acquisition costs and the complexity of analyses required by several applications, it becomes vital to share information before and after their integration. The current integrated systems, and even more so for future ones, require an increased integration and data sharing. A few examples of sharing requirements are inter-organizational systems, sectoral and corporate portals and strategic information exchanges.

Chapter written by Lin GINGRAS and François BERGERON.

When we look at what is being done now with GIS, we see the necessity of sharing and integrating even more. Several different methods and technologies currently prevent this integration; this cannot remain without jeopardizing the contribution of geographical IT (GIT) to corporate management applications. However, the need for location is very important and growing. Recent IT development has increased the need for this type of data.

In the following sections, we will review the main technological developments, which have, and will in the future, impact corporate management. We will then present some of the consequences of these technological developments on the characteristics of organizational applications and their impact on the development of GIS applications. Finally, we propose an evaluator of application opportunities for GISs, which could help companies in positioning their development.

3.2. Recent cases linking IT and GIS

3.2.1. *The recovery of Columbia*

A few hours after the unfortunate accident involving the space shuttle Columbia, DCIS (Defense Criminal Investigative Service) agents were busy documenting the location of debris from the shuttle using geomatic tools. A picture published in the Miami Herald on February 3, 2003, showed five people capturing different types of data, including geographic data captured by GPS (see Figure 3.1). This picture shows the use of a very sophisticated tool, but at the same time, it shows the sorry state of the American agency technology and explains the incredible delays in producing reports explaining the reasons for all sorts of catastrophes, American and foreign. The picture shows five people collecting information on paper, in a process that seems quite unclear for some of them (the two people on each end of the picture).

In a modern and up-to-date process, each agent would have a tablet computer or their PDA, which would integrate appropriate forms for collecting this type of information, and when collected, this information would be promptly transmitted digitally to a central database, which could be used to analyze the different components of a tragedy such as this one. Here is a public organization, overwhelmed by technology, and which, to provide a full report, will have to decrypt hundreds of documents with pen and paper. This information will then have to be recoded by remote people several days, if not several weeks later. How can we expect reports to be taken seriously and used efficiently? Organizations need applications that will process geographical data in an integrated way with their management systems. That is not what is happening here, even though the organization (NASA) is technologically advanced.

Figure 3.1. *Location of Columbia debris (source: Miami Herald, February 3, 2003)*

3.2.2. *Product management at Gillette*

Gillette from Boston, renowned for its razors, bought from Alien Technology 500 million radio-frequency identification labels (RFID) at the cost of 10 cents each, for managing an inventory of 60 million dollars. Antennas will be installed in its distribution center to identify stock movements. After delivery and "smart" shelf stocking, stock movements are transmitted to the retail store's inventory control system and the manager's PDA. The system will automatically thank the client for buying the product and will inform the manager when inventory is low. The labels can distinguish between sales and stolen products. These radio-frequency labels can enable a computer to identify any object, anywhere automatically and, in a certain way, make it possible for a product in transit to sense its own real environment. According to Richard Cantwell of Gillette, for a manufacturer, knowing where his products are is as important as knowing his bank balance [SCH 03].

Applications of this type are numerous and offer better control in many domains, for example, the field of drugs and medicine provided at no charge for physicians; all for 10 cents of wireless technology. The system will help Gillette save thousands of dollars in inventory maintenance costs and in their capacity to quickly respond to changing needs for their products. In addition, Gillette's resellers, such as Wal-Mart and Target, will also benefit from these control mechanisms to better manage their products and respond to their customers' needs in a more efficient way.

This type of application can, however, cause problems: the Italian group Benetton, famous for its sometimes not quite "politically correct" ads, announced the implementation of RFID tags made by Phillips for its Sisley product line. It now seems to be becoming cautious and withdrawing somewhat from this business strategy. Its website claims that "Benetton, which has always been a leader in technological innovation in the clothing sector, is currently analyzing RFID technology to evaluate its technical characteristics and emphasizes that no feasibility studies have yet been undertaken toward the possible industrial introduction of this technology. On completion of all studies on this matter, including careful analysis of potential implications relating to individual privacy, the company reserves the right to make the most appropriate decision to generate maximum value for its stakeholders and customers" [BEN 03].

3.2.3. *Trailer management at Schneider National*

The low cost of the implementation and high performance obtained by wireless tools and GIS, which can easily be integrated into logistics planning systems, made it possible for Schneider National to innovate by placing sensors in their 45,000 semi-trailers. The company has three trailers for each tractor and needs to be able to find out the status of each trailer, where it is and who is using it [MCC 02].

Schneider National uses these systems as a competitive weapon and as a way of serving its customers better. Its logistics systems are improved with such GIS tools and are widely supported by company employees, providing better efficiency and lower production costs. Technology has become an important lever at Schneider National for increasing the performance of the company's business plan. This leadership at Schneider National is followed by several other companies in the transportation sector, such as FedEx and Airborne Express.

3.3. IT developments

The following sections present a few of the most important trends in ITs applied to organizations: integrated applications, wireless networks, super portable computers such as tablets and PDAs, computer ubiquity, and newer, quicker and more powerful development technologies.

All these ITs are "disruptive technologies" creating discontinuity in the processes or in the use of a given technology by an organization or an individual. These technologies change the status quo, the way we do business by integrating increasingly smarter and quicker technologies, which are more geared toward user needs. These technologies eliminate the needs filled by a previous technology. The classic example is the disappearance of the Winchester disks from the market. They were big, expensive and powerful and were replaced by 5¼ inch diskettes that were not quicker, not more powerful, but they were smaller and much cheaper and go in networks that are more powerful and secure. Disruptive technologies have been changing whole wall spaces in IT in a continuous way since their beginning.

3.3.1. *Integrated business applications: ERP, CRM*

Since the mid 1990s, at least for American companies, we put aside home-grown development of applications targeting the inherent properties of companies to favor the implementation of integrated applications containing the best business practices of an economic sector or an industry. Standardized applications were implemented where personalization was at a minimum. This was done for different reasons, to decrease development costs, shorten the implementation cycle or benefit from the integration with these standardized applications among other reasons.

These enterprise systems, client management systems, specialized "business intelligence" systems and other applications of the same type all require great data sharing capabilities.

In addition, the trend was then to implement networked inter-company information systems with increased sharing of data inherent to the companies. One of the major components of data sharing is in the location of their origin, their geospatial context increasing their contribution to the performance of the partners of these company networks.

3.3.2. *Wireless networks*

In an article [RIS 03], it is estimated that 42% of American companies currently have wireless network applications and that more than 33% plan on implementing them during the next 6 months. Despite a wide range of standards, the growth of wireless applications is astounding, even if security, access and identification problems for people connecting to the network remain difficult. In a recent report, Gartner Dataquest group predicts a penetration of local wireless networks by 90% in 2007.

Wireless and mobile IT is currently at the center of many business innovations. It enables workers in different industries to transform the way they work in management of information and communications, as well as in their relationship with customers, suppliers, employees and business partners. From sales-force automation to customer relationship management, wireless computer networks are everywhere and create new needs for organizations and companies.

For example, new products using wireless technology make it possible to determine the location of people in buildings and activate appropriate management functions. In this way, a physician entering a patient's room with a tablet computer can instantly see the file of the patient in the room [SCH 02]. The same tool in retail environments could also provide information to the user based on his location.

In these contexts and in many others, the presence of wireless networks could become a disruptive technology, with the risk of annihilating traditional Ethernet networks and to modify the management processes of companies, which generates certain needs described later.

3.3.3. *Tablets and PDA*

The tools used in computer technology change quickly the way in which computer technology is presented to the user. Mobile, light and powerful, PDAs and tablets now make it possible to collect operational data at the source and to transmit information where it will be used in a business process, and so on and so forth.

These tools are deeply changing business processes making them more powerful, fast and accurate. There is no longer a need to transform the information digitally captured at the source; its processing is instant; because of wireless technology, the results are transmitted immediately where decisions are made. Managing and competing are no longer slow and centralized processes; we can increase reaction time and manage business networks ("edge management" and "edge computing"). Competition better beware!

3.3.4. *Quick development technologies*

Web time no longer allows companies and organizations to develop "at leisure" their business applications; competition is swift, instant and continually changing. Application development and business solution developers must follow this crazy rhythm.

Do we still have time to produce our own business applications? Many companies do not think so and use solutions already developed, such as enterprise systems. The cost of developing is too high, and the expertise is too rare to develop in-house. Another solution, group development by putting together expertise and the creation of business networks with their own solutions, makes it possible to reduce development cost and to maintain competitive capability.

Another approach leads to these results: outsourcing of computer services, for the development, as well as for operations and maintenance. It is actually a profound trend within the IT products and services industry. Companies no longer have the time to dedicate to fields where their expertise is not predominant; they let specialists provide computer support services.

3.3.5. *Ubiquitous information technology*

Ubiquitous IT [WEI 91] is a method for improving the use of IT by making several computers available in the physical environment while being invisible to the user. These applications are now a part of our environment without the technology itself being perceived as present. Many research studies are carried out in this perspective and result in new functions.

Ubiquitous computing occurs in social transactions and organizes and structures them each time these transactions occur. The challenge for the future is to make these ubiquitous applications even more present in current IT. In this field, GIS will have to integrate naturally to systems and applications, which are embedded in a wide range of situations and products.

Ubiquitous IT is largely the result of the introduction of wireless networks in organizations and the portability of computer tools. It changes methods in companies and organizations. In a large measure, it creates the needs and impacts described later. These needs and impacts constitute important development opportunities for management systems and GIS.

The trend toward ubiquitous computing is not only changing the way we acquire and treat the information, it is the beginning of a deep change in the way we access and use services and the beginning of the creation of new services, which only make sense in the way they are intimately present in the environment. It is the start of commerce, not of electronics or electronic business as is seen today, but the start of ubiquitous commerce and business. We must address the following question: how will business processes be structured in a world where the physical environment is full of services. The relationship between corporations and relations with customers and suppliers will be deeply modified [FAN02]. The potential change for business processes is important [FAN02]:

– location of the customer becomes the location of your company;

– a physical presence, wherever that may be, where your product or service is used, becomes a strategic necessity;

– mobile tools become the eyes and ears of remote service providers;

– mobile tools replace services associated with places such as travel agencies, banks, insurance firms and technical support.

Ubiquitous computing redefines the company–customer relationship. Companies and organizations, knowing the location and preferences of their customers and their providers, can become better able to respond to their needs and thus maintain better and more profitable business relations.

3.4. Consequences and impacts of IT developments on GIS

The following sections describe how current trends in IT have consequences and impacts on GIS needs and applications in organizations. In many cases, these impacts create immediate development needs and implementation of location organizational tools.

3.4.1. *Managing wireless*

Wireless networks, common intra- and inter-organization applications and ubiquitous computing deeply change the way we manage companies and organizations. Managing an organization is a group activity, and these new technologies make the notion of group very different from the one used to establish current management processes.

First, the group as we know it in current organizations no longer exists: common physical location, cable connected in an environment where social behaviors are well known is no longer one of the characteristics of modern organizations. We continually strive to improve organizational performance; we are getting closer to the organization's border. Ubiquitous computing supports activities in physically distributed environments, closer to the customer base and operations.

For example, decisions in an organization taken in geographically distributed groups generate application needs making it possible for participants to have contextual information, including the physical environment of group members.

Ubiquitous computing will enable the decision makers and knowledge workers to have much greater access to specialized data and processes. Ubiquitous computer technology will make it possible to access transactions when they occur from anywhere. Limitations of the office and traditional working hours are over. Decision makers can have access to complete information anywhere and anytime.

Ubiquitous computing provides a continuous follow up to goods and services and demand.

The value of ubiquitous computing for a decision maker clearly depends on the type of decision, but it seems that decision makers needing information from multiple people, processes and locations should greatly benefit from it. Having to maintain a workplace and the associated hours demands serious space and time constraints, which facilitate support services for office work. In addition, these constraints have negative effects on productivity; ubiquitous computing makes it possible to move office functions with the decision maker [DAV 02].

3.4.2. *Conscious context solutions*

Ubiquitous and mobile computing enables users to have access to the correct information at the right place and at the right time. Distractions become a problem if the user is concerned with his location, where he walks, drives and interacts with others. We have to minimize distractions for the users of ubiquitous computing. That is, when we have to integrate functions so that the system and the application can become context sensitive, the mobile computer knows the user environment and modifies his actions according to this knowledge. The context awareness almost always includes the notion of the user's physical location [SIE 02], his absolute and relative position, as well as his orientation. The objective here is to make the ubiquitous computing solution able to function in an assisting role using the contextual information to decrease the decider's attention needs.

For example, a notifying intelligent agent solution integrated into a PDA alerts a user when he reaches a certain minimal distance where a task that is on his to-do list is at hand: go to the fax area when a fax arrives, when he is close to a conference room and there is a meeting, etc.

3.4.3. *Information push*

Knowledge of the context through a computer solution, if user location is included, makes it possible to "push information" that is appropriate to the user's location [CHE 02]. For example, a card could enable a visitor to locate his position in a city and present information to him that is appropriate for his situation. This information push is actually most often triggered by the location knowledge of the system user.

The approach by push information can be compared with a more traditional pull-information approach of access to where a user is located. For example, a visitor could ask a geospatial system aware of the context and location about local tourist attractions. The approach by push information seems more appropriate in a context where the user interacts with his system, somewhat as with Web interaction. On the other hand, push-IT requires a degree of trust from the user that is greater than for a traditional pull-information system, because there is a certain measure of proactivity in the type of information transmitted.

In addition, these push-information systems based on context knowledge require more granular information on location than traditional systems, as well as an even greater capacity for instant response time. In all situations where the user is in information search mode, the capacity of push systems seems preferable by users of systems where the information is taken from a database [CHE 02].

Applications and business solutions using appropriate push information in a given context are potentially numerous for organizational management. However, there are even more of them in practice. The examples at the beginning of the chapter in transportation, inventory management and logistics are only the tip of the iceberg for many applications considered in the short term: for example, managing the processes in an agricultural company (fertilization, identification of appropriate actions based on location, information gathering after diagnosis), in personnel management and corporate activities, in customer and supplier relationship management, etc. In addition, we can reflect on the possibilities for a sales professional who meets with a customer, locating him, presenting information that is relevant to his situation, suggesting actions based on his environment, competition, etc. These tools are required to enable companies to compete more vigorously in many business contexts.

3.4.4. *User personalized information*

Applications and solutions, aware of the context, combined with relevant push-information systems for the user are the foundation for providing personalized information for the IT user, called I-centric communications computing [ARB 01].

Having to make choices or decisions, a system user may have preferences based on his location and environment. Alone in a place, he may be open to certain types of communications that he would not accept in another time or place. To target communications for the person, we must know where he is, hence adapted GIS.

The multiple communications tools, laptops, positioning systems and others are all technologies making communications targeted toward the individual possible: the presence of ubiquitous computing, flexible control methods for equipment in the environment (intelligent environments such as homes, cars, tablets or stores) and personal communications are becoming the fundamental elements of these systems.

The types of service in information personalization for an individual are those which define and maintain contexts tailored to the preferences of a single individual in his personal way of interacting with a communications system. On the basis of preference and capability profiles and the information linked to location of a person, a user can be offered individualized services according to his demands. In addition, intelligent environments that identify spatial and temporal data personalize the interaction.

3.4.5. *Digital artifacts in embedded tools and objects*

Personalized information for the user requires the creation of embedded digital artifacts and tools in different products in different places. This presence of artifacts provides the availability of a product or service to locate it, personalize it based on a present or remote user and to follow its performance. These tools must almost always be able to determine the location because, in most cases, the user uses location to determine the relevance of the product or service proposed. GIS becomes an important part of these tools.

This type of tool requires applications that either move with the user or are present in the user's environment. Among them, there are many applications for locating the person in his environment and the organizational performance they carry. For example, sensors in a building can locate who is there and can validate their presence. It is a wonderful tool for security of the person and premises. However, there could obviously be certain social constraints: does a person want to continuously be identified and have his behavior automatically captured? Certainly not in every room in the building! In addition, who has this information? Ubiquitous computing will need to respond to some of these questions before becoming more prevalent [BAN 02].

3.4.6. *The other dimensions: application structure and security applications*

Current trends in ITs have an effect on other GIS dimensions: applications locating the source and security applications are two of the most important.

The development of modern applications using ubiquitous computing requires more powerful information structures that include location of the information source, especially if this information comes from the Internet [KOR 02]. We want to be able to physically locate an object, its orientation, dimension, direction, the moment, etc. In addition, for administrative Internet applications, the capacity to locate the contributor, product or service by the user of a service on the Web is an important element of the trust he has in it, a predominant element in the performance in a transactional website.

Location is also a vital component for applications in the field of corporate security. In the current context the demand for security, identity authentication, information access control and transaction authorization, developing applications to locate products, customers or suppliers for security control purposes from corporate operations is becoming strategic.

3.5. Potential gains and risks with GIS

Organizations must make strategic choices when implementing and developing their computer applications. To our knowledge, currently, there are no tools enabling corporate decision makers to take into consideration development risks and strategic advantages with GIS. The following sections present a frame of reference on the introduction of these technologies in the corporation.

Wireless technology and ubiquitous GIS applications offer great benefit potential for organizations. These benefits are grouped into three parts: operational, managerial and strategic. However, as with any new technology, there are risks that should not be ignored.

3.5.1. *Operational benefits*

GIS provides the company using it with several types of operational benefits. They increase the quantity of data produced during a collection operations among others. They also enable better quality of information; as it is more precise during collection, it can be validated immediately by the computer system or the computer chip supporting the ubiquitous tool. It is also error-free, from computer to computer for greater security and data integrity. Delays in information transmission are also improved and completely optimized. In fact, data can be transmitted as soon as it is captured with no delay.

Operation costs of the corporate activity supported by a ubiquitous system can also decrease. That is the case in companies using GIS to manage their truck fleets.

3.5.2. *Managerial benefits*

A second category of advantages involves the improvement of managerial control. The use of GIS offers benefits when it is integrated into the operation of the company. Projects can bring interesting benefits insofar as they were already evaluated on the direct costs that they incur, as well as on indirect costs and compared with monetary benefits that can be quantified. As for the intangible benefits that they generate, they were used too much in the past as the single justification for projects which obviously were not profitable and should never have been launched. Once this project evaluation step is done, certain GIS applications help in managing a product's lifecycle, with their assembly in a plant and with subsequent steps of the value chain, with the distribution of the product and even through the CRM function. The Gillette case presented earlier is an obvious example. They can also be useful for personnel management, as in the example of geomatic systems in semi-trailers.

3.5.3. *Strategic advantages*

Strategic advantages with GIS systems are not automatic. However, when applications are correctly identified, they enable the company to better position itself on the market, increase its market share, increase its customer base, offer new products, accelerate its growth and increase its profitability. Ultimately, any large GIS project as with any important IT project must contribute to the corporation's bottom line. This is possible if the project makes it possible to offer a new service to customers, to support suppliers more efficiently, to eliminate competition by a smart choice in applications or by preventing the arrival of new competitors on the market. The integration of GIS information produced by the company's internal and external value chain by establishing privileged communications among customers, suppliers, distributors or government agencies will facilitate the generation of these strategic advantages.

3.6. The importance of strategic harmonization for project success

The strategic harmonization of GIS projects is vital to their success. By strategic harmonization, we mean the alignment of corporate strategy and IT strategy and the alignment of corporate structure with the IT structure. It is only when these dimensions are aligned that a company can hope to derive concrete benefits from GIS investments.

3.6.1. *Business strategy*

3.6.1.1. *What we mean by strategy*

There are several different definitions of business strategy. There is the concept of desired versus actual strategy [MIN 79]; in other words, the strategy that we wish to achieve compared with the actual strategy that happens in the real world. There is a market approach strategy developed by Porter [POR 90], which cites that a company has a choice between three possible strategies: costs, differentiation or niche. We now know that these strategies are not exclusive. The following are the types of strategic: Developer, Analyst, Defender and Reactive [MIL 78]. We also find a hierarchy of strategies: vision, general (corporate) strategy, operational (divisions) and tactical (operational units) strategies [JOH 99], etc. All these strategy classifications have their use, but we prefer strategic orientation activities [VEN 89], which has the advantage of explaining the strategic actions carried out by the company.

3.6.1.2. *The intensity of strategic actions*

Based on the concept of strategic actions intensity, it is possible for a company to execute different types of actions some of which particularly lend themselves to the concept of strategic harmonization.

The four types of actions are as follows: proactive, analyst, defender and futurist. For each of these types, the company can analyze its position on the market, make appropriate choices and carry out specific actions. The intensity of strategic actions has more to do with the effort by the company to choose and implement its strategy than with the type of action itself. A company with a strong strategy carries out several strategic studies and implements the results of its choices, whereas a company with a weak strategy does not carry out many studies and implements few actions directly resulting from these choices. To obtain financial performance that is superior to competition, a company must have a strong strategy. In a GIS context, the company that will implement a larger number of GIS applications will have a stronger strategy and should gain more benefits. That is the case with Schneider National presented earlier, which remains a leader in its field with many attempts with GIS technologies.

3.6.2. *The GIS strategy*

A corporate IT strategy is made up of two main dimensions: (i) environmental IT watch, including GIS applications, and (ii) the strategic use of these applications.

3.6.2.1. *Environmental IT watch and wireless technology*

The environmental IT watch dimension involves the actions that the company carries out to understand available GIS IT, and how the industry uses them, along with the competition. The environmental watch must be a permanent activity. For example, an organization must know how competitors use recent GIS technologies: Are they using wireless technology? Can we locate the customer visiting our competitor via radio to help him find what he is looking for as is done in some American superstores? Technological watch ensures corporate survival against competitors using technological levers such as GIS.

3.6.2.2. *Strategic use of GIS applications*

The strategic use of GIS applications mainly involves the profitability and return aspects of applications for the company. It therefore includes, among others, the use of IT to decrease production costs, increase the company's productivity, respect delivery dates promised to customers and increase the bottom line of the company. For example, facing intense competition, Dillard's chain of department stores [DUV 03] was able to use GIS applications to accurately identify the location of its sales per store instead of regionally, thus increase their stock turnover and ensure better financial performance.

In order for the projects to benefit the company, the IT strategy must be strong. It must align with the corporate strategy, while being the work tool of strategic analysts or the IT tool used to achieve day-to-day corporate strategic actions. IT only benefits the company when the IT strategy is strong and if it supports a strong corporate strategy. Companies with a weak IT strategy invest at a loss in GIS applications and this situation must absolutely be avoided.

3.6.3. *Corporate structure*

For a small- to medium-sized company, the quality of corporate structure corresponds to its level of task formalization, worker specialization and its number of hierarchical levels. For service companies, we add the level of staff professionalization, whereas for the manufacturing company, the level of centralization is important. Companies with clearly defined tasks, where workers are specialized and with at least three hierarchical levels (i.e. the CEO is Level 1, Level 2 is where we find specialized staff and Level 3, the workers), derive greater benefit from IT and GIS than companies not meeting these criteria. The quality of the corporate structure is therefore important.

3.6.4. *GIS environmental structure*

IT and GIS application environmental structure is based on two dimensions: the first one involves IT planning and control and the second involves IT acquisition and implementation. This corresponds to the structure of the process of widened IT management. We should note that we are not going to analyze structure from IT, in other words computer architecture, applications and data, nor from the organizational angle including a centralized, distributed or not, IT structure, because these configurations are secondary in importance in relation to the need for a structured process of IT and GIS resource management.

3.6.4.1. *Planning and control activities*

Planning and control activities involve a series of elements inherent to IT but seen from the perspective of a corporate executive. This dimension involves elements such as understanding of current technological products, controlling any project involving IT or GIS, the capability of identifying IT needs and finding GIS solutions, planning of any IT project, including GIS projects, in relation with the business objectives of the company and developing a technological culture.

3.6.4.2. *The activities of acquisition and implementation*

The acquisition and control activities of technologies involve the use of a structured approach to project evaluation. This approach uses specific criteria of evaluation and financial tools analysis. It includes a prior evaluation of the impacts of new technology on company personnel and on organizational changes that it will bring. Finally, it reports the contributions of this technology for helping companies reach their objectives and the problems that it could generate.

It is therefore important for a successful GIS application to ensure that not only is the corporate structure sufficiently developed but also that the IT environment structure is accurately managed. It will be difficult for an unstructured company to know how to adequately manage IT. It is often in these cases that computer applications will not be profitable for the company. In less structured environments, we cannot implement a solution quickly because that obviously requires important organizational change.

This harmonization of structures, paired with the harmonization of strategies, as well as the crossed strategy-structure harmonization, is the keystone of profitable computer applications.

3.6.5. *Corporate performance and strategic harmonization*

Strategic harmonization benefits are expressed by increased corporate performance. This performance is based on two dimensions: growth and profitability. Growth includes an increase in sales, the development of new products and an increase in market share, whereas profitability is measured by an increase in the rate of return over sales, capital performance and profit per action.

Harmonization of strategies and structures enables IT projects to maximize the chances for success measured by corporate performance. In fact, this harmonization is very important because we have observed that companies with low harmonization are less profitable companies. GIS projects, as with all other IT projects, must have the same rigorous evaluation, the same careful management and the same strategic alignment as other IT projects. These requirements are worth it because, in the long term, they will allow people to use GIS applications more efficiently. The organizational performance criterion is the ultimate evaluation of success. This grid applies to the large corporation and to the small- to medium-sized company.

3.7. An indicator of GIS application opportunity

GIS applications are very similar to the concept of competitive applications. In that capacity, different approaches can be used to identify application opportunities. In fact, Porter's value chain approach is the most appropriate for manufacturing companies, and the Wiseman model provides better results for service companies [BER 91]. We propose a GIS project evaluation grid for these approaches called UBIGIS to better select promising projects for a company, and thus decrease risks and increase performance.

3.7.1. *The "UBIGIS" grid to evaluate GIS application opportunities*

The UBIGIS grid (Figure 3.2) enables us to evaluate GIS application opportunities in a corporation. It is made up of three main parts: benefits, strategy and structure. The benefits must be positive, the strategy must maximize the intensity of strategic actions and structure must be developed enough to support the strategy.

Benefits

Operational benefits:

Increase in the quantity of relevant data
Improved quality of data (precise, up to date, complete)
Validation of data at collection
Error-free transmission
More precise location
Improved mapping
Shorter transmission delays
Better data security
Decrease in involvement delays in the field
Greater integrity of data
Decrease in direct production costs (labor and parts)

Managerial benefits

Decrease in indirect production costs
Quicker intervention and decrease in losses
Decrease in total cost of production (e.g. less product losses)
More efficient use (follow up and control) of human resources in the field
Expansion of the CRM database (intangible)
Better ERP implementation in the value chain (intangible)

Strategic benefits and corporate performance

Increase in customer base
Increased satisfaction of local client base

Increase in sales linked to product/service
Company growth (sales)
Increase in profits
Decrease in costs compared with direct competitors
Gaining competitive advantage

Strategy

Business strategy (strategic actions)
Proactive (new products/markets)
Analytical (extrapolation of trends)
Defensive (maximum decrease in costs to retain a market)
Future (forecasting future actions)

GIS strategy
Environmental IT and GIS watch activities
Strategic use of IT and GIS applications (concrete contributions)

Structure

Corporate structure
Formalization (documentation of tasks, procedures)
Administrative intensity (manager/employee ratio)
Professionalization (professional/employee ratio)
Specialization (distinct tasks)
Differentiation (number of hierarchical levels (for small-medium sized companies))

GIS structure
Planning and control (planning/control systems structure)
Acquisition and implementation (hardware, software, networks)

Figure 3.2. *UBIGIS grid*

3.8. Conclusion

GIS applications represent a very powerful technological breakthrough for companies and organizations. As Baltazar says [BAL03], "the investments in GIS technologies today could help in the creation of new companies and open new markets, while decreasing costs by helping corporate management obtain more from the organization's data and personnel".

The examples of applications presented enable us to consider important changes in corporate structure and their operations. New technologies and GIS applications will soon flood the companies and they will lead to great interest among managers and users. Because history has a tendency to repeat itself, companies could let themselves be convinced of the usefulness of these new applications, and many projects will be launched on the faith of optimistic recommendations from developers of this technology. In this group, we find systems analysts, hardware, software and wireless network vendors, management and IT consultants.

The information systems discipline now has enough history to avoid falling into the same pattern: no IT or GIS project, whether ubiquitous or not, wireless or not, should be started without a complete evaluation and quantification in cost of tangible and observable benefits on performance of the company, as observed by its growth and net profit. Finally, these projects must also be implemented in a favorable organizational context and supported by upper management. In this way, GIS applications will have the best chances of success while decreasing the risks associated.

3.9. Bibliography

[ARB 01] S. ARBANOWSKI, S. AN DER MEER, S. STEGLICH, R. POPESCU-ZELETIN, "The human communication space: Towards I-centric communications", *Personal and Ubiquitous Computing*, vol. 5, Springer-Verlag, 2001, 34-37.

[BAL 03] H. BALTAZAR, "Hurdles to GIS Lowering", *eWeek*, 47-48, 2003.

[BAN 02] G. BANAVAR, A. BERNSTEIN, "Software infrastructure and design challenges for ubiquitous computing applications", *Communications of the ACM*, 92-96, 2002.

[BEN 03] Benetton Group, "Benetton: No microchips present in garments on sale" press release document, Benetton Group, Available at: www.benetton.com/press, 2003.

[BER 91] F. BERGERON, C. BUTEAU, L. RAYMOND, "Identification of strategic information systems opportunities", *MIS Quarterly*, vol. 15, 89-103, 1991.

[BER 92] F. BERGERON, L. RAYMOND, "The advantages of electronic data interchange", *DATABASE*, vol. 23, 19-31, 1992.

[BER 95] F. BERGERON, L. RAYMOND, "The contribution of IT to the bottom line: A contingency perspective of strategic dimensions", *Proceedings of the 16th International Conference on Information Systems, Amsterdam*, 167-181, 1995.

[BER 01] F. BERGERON, L. RAYMOND, S. RIVARD, "Fit in strategic information technology management research: An empirical comparison of perspectives", *Omega*, vol. 29, 125-142, 2001.

[CHE 02] K. CHEVERST, K. MITCHELL, N. DAVIES, "Exploring context-aware information push", *Personal and Ubiquitous Computing*, vol. 6, Springer-Verlag, 276-281, 2002.

[CRO 01] A-M. CROTEAU, F. BERGERON, "An information technology trilogy: Business strategy, technological deployment and organization performance" *Journal of Strategic Information Systems*, vol. 20, 77-99, 2001.

[DAV 02] G.B. DAVIS, "Anytime/anyplace computing and the future of knowledge work", *Communications of the ACM*, 67-73, 2002.

[DUV 03] M. DUVALL, "Dillard's: The big fix", *Baseline*, Available at: www.baselinemag.com, 2003.

[FAN 02] A. FANO, A. GERSHMAN, "The future of business services in the age of ubiquitous computing", *Communications of the ACM,* 83-87, 2002.

[HEN 93] J.C. HENDERSON, N. VENKATRAMAN, "Strategic alignment: Leveraging information technology for transforming organizations", *IBM Systems Journal*, vol. 32, 4-16, 1993.

[JOH 99] G. JOHNSON, K. SCHOLES, *Exploring Corporate Strategy*, 5th Edn, Prentice Hall, Europe, 1999.

[KOR 02] M. KORKEA-AHO, H. TANG, R. SULONEN, "Expressing location for applications on the internet", *Personal and Ubiquitous Computing*, vol. 6, 329-333, 2002.

[MCC 02] J. MCCARTHY, "The wireless road taken", *Infoworld*, 38, 2002.

[MIL 78] R.E. MILES, C.C. SNOW, *Organizational Strategy, Structure, Process*, McGraw-Hill, New York, 1978.

[MIL 87] D. MILLER, "Strategy making and structure: Analysis and implications for performance", *Academy of Management Journal*, 30, 7-32, 1987.

[MIN 79] H. MINTZBERG, *The Structuring of Organizations*, Prentice-Hall, Englewood Cliffs, NJ, 1979.

[POR 90] M.E. PORTER, *Competitive Advantage*, The Free Press, New York, 1990.

[RIS 03] O. RIST, "Untethering the enterprise", *Infoworld*, 35-40, 2003.

[SCH 02] E. SCHWARTZ, "Wi-Fi intelligence gets boost", *Infoworld*, 16, 2002.

[SCH 03] E. SCHWARTZ, "RFID about to explode", *Infoworld*, 28, 2003.

[SIE 02] D.P. SIEWIOREK, "New frontiers of application design", *Communications of the ACM*, 79-82, 2002.

[VEN 89] N. VENKATRAMAN, "The concept of fit in strategy research: Toward verbal and statistical correspondence", *Academy of Management Review*, vol. 14, 423-444, 1999.

[WEI 91] M. WEISER, "The computer for the 21st century", *Scientific American*, 94-104, 1991.

PART 2

GIS and its Place Within Organizations

Chapter 4

The Role and Value of GIS in Organizations

4.1. Introduction

In the context of the new economy based on knowledge, information technologies (IT) constitute vital tools. Among this new IT GIS are to support our understanding of the territory, as well as the physical and human phenomena found on it. For the last 20 years, these technologies have been evolving and systematically integrating innovations from the general IT evolution trend. GIS, which can now be implemented to understand and manage the territory, is very diverse: GIS with GPS satellite receivers to locate vehicles in real time, GIS integrating satellite images to follow the evolution of forests, GIS to analyze digital ground models (3D) for landscape preservation, GIS to visualize or simulate the effects of natural catastrophes, web-based GIS to broadcast urban maps to the public, GIS for real-time management of underground conduits in cities, etc. The potential impact of these innovative tools on corporate performance is obviously important.

Chapter written by Claude CARON.

Basically, GISs exist to bring information to decision makers; therefore, they have a "value" for organizations. Nevertheless, for many executives and companies, this value is often difficult to define and measure. Sometimes, it is not easy to determine the exact role that a GIS will play within an organization beforehand. Because of its uncertain role, it is certainly not easy to choose the appropriate measure of efficiency for this role.

Despite the attraction for web-based GISs, the initial interest generated by the novelty of GIS has now faded. In summary, the role played by GIS has gone from "new toy" to a technological tool that is increasingly accessible and integrated into mainstream IT. In terms of the value of GIS, the justification now required by financiers concerning investments necessitates that the measure of their value be transferred progressively from an art to a science. In this evolutionary context, where knowledge economy organizations already invested massive sums to implement GISs, a better determination of the value and a better understanding of the exact role played by these systems are vital.

In the following sections, we will first address the social context in which GISs fit. This is essential to understand their rational and symbolic role better. Examples of typical roles will help us highlight a general perspective of evolution of the GIS role. We will then discuss GIS value from a philosophical and economical point of view. Next, we will analyze the qualitative and quantitative evaluation approaches. Finally, we will connect the concepts of role and value. We will then be able to understand how the role and value specifically attributed to a GIS can greatly influence the resulting technical and organizational environments.

4.2. Social context of GIS: images, symbols and rationalities

To truly understand the role of GIS, we must first define the organization and social context in which these ITs play their role. In fact, the social context in which these technologies evolve constitutes the stage in which we, as director, attribute their major role.

4.2.1. *GIS: a rational role*

In our Western societies, decision making and acquisition of information has become fundamentally important. This importance is concretely shown in all tasks and organizational structures where GISs often constitute an important lever for reaching the goals set. As said by Martinet [MAR 91, p. 12]:

> *[...] building of information can be considered as more important than the information itself, just like making a decision is more important than the results. Since modern Western civilization made intelligent choice and rationality into major values, it is not surprising that decision and information are places and challenges of competence demonstration and quest for legitimacy for "decision makers" as well as for organizations.*

The classical theory of decision stipulates that "decision makers who know how to make decisions are those requiring information, collecting it and acting on it. That is the behavior considered appropriate in terms of decision making" [FEL 91, p. 266]. This symbolic vision of decision making is similar to the necessity of displaying an *image* of success. In fact, as Feldmann and March mention [FEL 91, pp. 264–265]: *these symbols of competence are also symbols of social efficiency, which justifies them in part.* Actually, the use of symbols of competence, as with the efficient use of GI, is, in part, because it is usually easier to consider a decision as *rational* in terms of the process followed (form) than in terms of its substance (depth) [FEL 91, MAR 91]. Very often, the process of decision making itself is more important than the actual decision made, since this process offers:

> *a base for applying social values, to be authoritative, and display appropriate behavior in relation to one of the central ideological creations of modern Western civilization: the concept of intelligent choice. Bureaucratic organizations are buildings built on the*

> *idea of rationality. The cornerstones of rationality are values for decision making [...] [FEL 91, p. 264].*

But what is rationality? From a decision making and informational point of view, the so-called *classic rationality* stipulates that *humans make choices* [MAR 76]. This traditional idea or belief is deeply rooted in our Western society. It implies that the decision-making process can be improved and accelerated by using *techniques* (methods, software, etc.). They contribute to finding other alternatives to improve the quality of the information and the efficiency of the analysis for evaluating alternatives. Then, when it is done correctly, the choices are made by evaluating the alternatives in relation to the objectives and on the basis of information that is actually available.

Consciously or not, GIS rationality fits into this creed. It appears to be deeply rooted in our culture and has conditioned a large part of our contemporary Western history, as well as our current interpretation of this history. These ideas remain well rooted in our theories involving decision making [MAR 76]. In fact, *classic rationality* basically stipulates that *technology contributes to social progress*. Guilhon [GUI 93] summarizes implicit and rational causal relations in terms of technological development (Figure 4.1).

Research & Development	→	**Technological transformations**	→	**Economic growth**	→	**Social progress**

Figure 4.1. *Causal relations involving technological development [GUI 93, p. 29]*

This *hierarchy of goals* is often automatically accepted in our society. Nevertheless, there are a very few empirical studies to this day making it possible to systematically demonstrate the actual and positive impact of IT on corporate performance.

In the same vein, the widely known idea in management is that, with training the introduction of IT and optimization and simulation techniques, it is possible to rationalize and improve decision making, production processes, and from that, generate productivity gains in organizations [GUI 93] (Figure 4.2).

Technological developments	→	**Production process improvement**	→	**Gains in productivity**

Figure 4.2. *Traditional usefulness of technologies [GUI 93, p. 19]*

Based on a *rational* corporate vision, strategies are often considered as pre-established, presuming that objectives and instructions are clear, that decision makers know what they want and finally, that what they want is consistent, stable and not ambiguous [BAI 91]. General directions come before global and consistent actions, and appear exogenous, coming from a legitimate authority. In this context, IT is most often considered as able to increase corporate performance to reach the strategic objectives of the organization.

4.2.2. *GIS: a symbolic role*

Clearly, GISs do not only play a rational role. For dozens of years [BOO 71], the Western social context in which our different organizations (companies, ministries, societies, cities, etc.) evolved was characterized predominantly by the importance of *images*. The fundamental ideology is very real and omnipresent, although it may seem somewhat distorted. It is the one that influences most of us in a subtle way with the *dream of welfare and success for all.* This ideological context from North America also supports *equal chances for all* and *individual freedom.* Everybody gets in the race where the *law of the strongest* is the basic rule. As is well summarized by Marchand [MAR 88]:

> *Step into the race, play the game and you will succeed! [...] May the best survive: we live in a world where there is only room for winners and where the principle of productivity is the ultimate rule and the universal unit measure. [...] To be free, you only need*

> *to be in the race; to succeed, you only need to win the race [MAR 88, pp. 79–80].*

However, if everyone is in the race, who determines the winners? Answer: everyone except ourselves. Success, in fact, is only based on an *image of success* and victory. And this image, *we can only find in other people's judgment: in the recognition, admiration or even adulation that winners inevitably have* [MAR 88, p. 81]. In this way, broadcasting an *image of success* is more important than the success itself. To reach this objective of *success*, a very powerful social way is proposed implicitly to organizations: *technology* (where GIS constitutes a particular configuration). Since it constitutes a *way*, a *strategic competitive weapon* [BLI 89] to symbolically reach *success*, it in turn becomes a *symbol*:

> *Collectively, [the technical object] is a symbol, to the fullest extent of the term: a sign of recognition shared by all supporters of scientific and technical progress, a creed which all social stakeholders share. [...] This system of objects symbolizes our desire for modernity; our consensus to build through technical progress the optimism of happiness in a perfect society. [...] Production and social distribution of new technologies thus resembles magic [...][SCA 89, pp. 109-110].*

We must admit that most of us believe in the advantages of progress, specifically progress through IT (including GIS). This technological optimism, which characterized 19th century *positivism*, is a belief that is revived at every new miracle that science and technology brings to us, which is certainly also the case with GIS.

4.2.3. *The role of GIS: multiple rationalities*

Generally, according to the rational and prescriptive vision, what is not rational is a problem that has to be fixed or a bias that should be eliminated [PIC 90]. On the other hand, we now know that GISs often evolve in organizational contexts where ambiguity, complexity and uncertainty are pervasive [CAR 00]. In a more contemporary way, contrary to the *prescriptive* vision of decision making, a somewhat descriptive vision of the process has emerged. According to this vision, we consider that a decision is always rational, but that there are *multiple rationalities*, one can always explain a behavior. In this way, to explain the rational role of GIS in more detail, different qualifiers can be associated to the term rationality: limited rationality, contextual rationality, game rationality, adaptive rationality, etc.

In a general way, considering human behaviors as being *coherent* in relation to certain forms of rationalities has widely contributed to improving our understanding of human decision making. Behaviors are now considered as not only understandable but also, in a large measure, intelligent.

Here are a few important general characteristics of the social context in which GISs play different roles. Activities, decisions, information and IT fit in a society where the *image of success* is fundamental, the *symbols* are powerful and pervasive and *rationalities* are numerous.

4.2.4. *Multiple GIS rationalities: a few typical roles*

In the light of the rational and symbolic approaches presented, we now propose to put GIS in different scenarios to concretely understand the notions addressed. The typical roles presented here and inspired [ROC 96] are deliberately well constructed. They represent fake situations to illustrate the wide range of roles played by GIS.

Trend role: three high-tech firms want to open offices in the area. City "A" wants these companies to set up offices on its territory and faces fierce competition with neighboring city "B". However, city "A"'s image is quite dull, whereas its competitor's is very dynamic and modern; this more modern image comes from the fact that city "B" has recently implemented GISs. City "A" decides to follow the lead and invest in the updating of its image by acquiring the latest and greatest in GIS technology. In this context, a GIS implementation is initiated by a driving effect or trend. The organization decides to implement a GIS because another similar organization already has it.

Technological showcase role: in certain cases, the goal is not to "keep up with the Joneses" but to distinguish themselves. It is the case of a regional economic development agency that wants to maintain an image enabling it to distinguish itself from neighboring development agencies. This agency discovers the innovative concept of web-based GIS and decides to implement a promotional Website for the region. They of course make sure they publicize the operation to showcase the distinctive character of its sector. In this specific context, GIS constitutes a *showcase*, a symbol distinguishing the agency from its competitors.

Symbolic role of the information-decision link: the ministry in charge of land management is a governmental bureaucracy that is well established. In a budget-restriction period, it must constantly justify its scope and the resources it uses. In practice, even if the decisions are becoming more complex, an increasing number of decision makers are at loose ends. Information collection and decision making sometimes do not connect in the organization. An executive even admitted the other day: *In fact, we often make different types of decisions here, without always knowing what the global purpose is; what concerns me is showing that I use the technologies available to me (GIS) and that I always take the time to analyze all possible scenarios before making a decision. I have to always show a rational and professional image!* The emphasis is in the form of the process (the methodology) instead of the problem itself. GIS is perceived as a way of reinforcing the information-decision links, as well as rational and symbolic dimensions of decision making.

Operational role: a city's public works department must internally ensure the maintenance and operation of its water supply system. Two civil engineers ensure the continuity and knowledge of information relating to this system. One of them passes away and the other one will soon retire. The city then has to face the risk of losing the only up-to-date information involving the system: the knowledge of the last civil engineer. A solution must quickly be found to face this problem, because the city must continue to be able to maintain its water supply system. It wants to make the system even better by obtaining real-time information about the system, which would enable it to know which gates to close and which residents are affected during breakdowns. In this operational context, the role of GIS is to enable maintenance and updating of infrastructure data involving the system.

Legal support role: a real-estate company launches a large project involving the construction of residential houses and apartment buildings. Current city planning rules in the designated zone are particularly complex. They involve rigorous control of the location of constructions and their architectural characteristics. It is important for the company to verify the appropriateness of chosen planning scenarios with the help of simulations and analyses done by GIS. In addition, this same information system will make it possible to ensure the appropriateness of these new houses in relation to regulatory and legal constraints down the road.

Persuasion role: a real-estate promoter has a large piece of land to develop on the west side of the city. He wants the new highway to go through his property for two main reasons: (1) expropriation would be very lucrative for him and (2) the remaining plot would go up in value. He is soon meeting city counselors to show them the maps he had drawn up with the help of a GIS to convince them of the best location for the new highway. In this context, the GIS is used for its potentially efficient GI broadcasting in rich and various forms for persuasion.

Planning support role: an organization involved in regional forest development has the objective of planning the actions to occur to landscape a large section of the regional forest. This type of planning must be carried out in coordination and discussion with several people involved, each with their own concerns: preservation of the landscape, forest exploitation, tourist and recreational development, wildlife conservation, etc. In a decision-making context, for sustainable development planning, a GIS is used to support such tasks as scripting, simulation, spatial analysis and production of thematic maps.

Prospective or exploratory role: a private transportation company, aware of its precarious position in the market, goes in search of new technologies enabling it to gain a competitive advantage over its regional competition. It has heard that similar companies in other countries installed GPS receivers in their fleet to position their vehicles in real time over a GIS installed at the main office. It decides to install these receivers in its trucks and integrates localization data in a GIS. This enables the company to efficiently follow the route of its trucks and inform customers of the exact location of their orders. This company is taking a calculating risk by using GIS technologies in an innovative way in its particular field.

4.3. Evolution of the GIS role

GIS technologies have evolved considerably in the last decades; this evolution is done by constantly converging toward the vast field of IT. In this way, and in a constantly changing society, the role of GISs is also very evolutionary.

4.3.1. *Toward more diversified roles*

For the last two decades, organizations in different fields linked to territory have traditionally used GISs for the management of natural resources. Their use has diversified recently toward new business sectors (customer management, geomarketing, tracking of packages (UPS, FedEx, etc.) or fleets of trucks, etc.). In addition, this use is moving in a big way toward “mass public” applications (light GPS receivers with mapping displays for leisure, OnStar navigation in vehicles, static (e.g. MapBlast) or dynamic (e.g. personalized bicycle path itineraries) Internet mapping, personal digital assistants (PDA) with GIS showing pizzerias at less than 500 feet, etc. For a greater description of the perspective of GIS technological evolution, the reader should refer to Chapter 10 in this book.

In summary, through the observation of several organizations where GISs have been implemented in the last few years, we can make the following main assessments:

– First, geomatic IT evolve toward lighter versions (e.g. mapping on the Web) and mobile configurations (e.g. palmtop or laptop with GPS receiver and cartographical display) [VAN 01].

– Second, the business sectors where IT is used and the types of users are becoming diversified [SON 99]; mainly made up of land managers, the customer base of GIS technologies now groups people practicing multiple sports and leisure activities.

4.3.2. *Toward higher decision levels*

Numerous studies were carried out involving the link between IT and corporate strategic objectives [CAR 97]. Several authors emphasize that IT themselves have very little advantage for organizations. Actual benefits are the result of the capacity of organizations to plan their deployment and use by connecting them to corporate business objectives. In this way, the fact that IT must align with business strategies is a generally accepted assertion. This attraction for strategic IT planning has picked up during the 1980s and has not stopped growing.

In terms of GIS, in particular, recent studies tend to demonstrate that these systems were traditionally only slightly linked to strategic levels of organizations [CAR 97]. According to a study carried out at least more than 200 GIS projects in Canada for a 10-year period, the main objective of 63% of the projects studied was of tactical or operational nature and not strategic [CAR 01].

In the light of the latest innovations in the field of GIS (light and mobile configurations) and of new organizations offering new roles for these technologies, we notice that these systems currently offer a competitive strategic advantage. These new technologies also have a strategic advantage by enabling an organization to differentiate itself.

Is this situation short term, i.e. only linked to certain recent technological innovations integrated in GISs? Or is it a megatrend that will support the link between new GIS technologies and strategic concerns? Is a link between GIS and business strategies possible and even desirable in all cases? Regardless, it is a potential evolution that will certainly be interesting to observe during the next few years.

4.3.3. *A succession of different perspectives*

With hindsight, in terms of the development of GIS technologies, it is now possible to bring out three complementary perspectives in succession [CAM 97]: technological determinism, executive rationalism (one best way) and social interactionism [ROD 02]:

– *Technological determinism:* on the basis of this perspective, GIS technologies are beneficial and inevitably generate progress. The perspective and methodology are mainly technical and bound to be successful insofar as the resulting GIS is optimal. GIS literature was widely dominated by this perspective, especially in the early 1990s.

– *Executive rationalism:* the introduction of GIS in an organization does not involve only the technical aspects. Its success is not automatic and depends on following a rational methodology (cookbook method), based on a development cycle (waterfall approach). The unique but adaptable "recipe" takes into consideration factors influencing the mechanistic development of the project. This perspective is widely covered in GIS development literature.

– *Social interactionism:* with this perspective, the implementation of GIS is a single interactive process between technologies and a given organization. The same GIS technologies, implemented in organizations and following similar methods, can provide very different results. In this way, *other forms of knowledge such as experience, prior beliefs, values and rumors may become as important. Because of this, morality, ethics and even tactical considerations can have more significance than traditional notions of rationality and logic* [CAM 97].

Few research studies today are attempting to bring GIS development projects from the second to the third perspective. Nevertheless, since the mid 1990s, initiatives have been in place to do that and are causing a fundamental questioning, both from the philosophical and practical (economical) point of view, involving ways to determine the value of these technical and organizational solutions.

4.4. The value of GIS[1]

The definition of the concept of *value* still constitutes a major problem today. Over time, from Plato and Aristotle to economists from the information age, this concept was defined according to different and various perspectives without ever reaching a real consensus. The various considerations for this concept, different in philosophical, economical and political aspects notably, make it very subjective and variable in practice.

In the literature, the term *value* goes back to the relative and non-absolute range. The measure of value can be considered as the relation between the usefulness of objects and the needs (often associated to desires) that these objects can satisfy. Regardless of the exact meaning of the word value, there is no inherent generic unit to express the measure of this value, as there is, for example, for the measure of physical phenomena (distance, angles, geographical location, etc.).

4.4.1. *Value: philosophical or economical concept*

Axiology is the science of values of estimation and assessment. These concepts are usually linked to the perception of subjects. The value is thus a subjective concept, a relation of people's estimation of objects. Measuring the value of an object involves knowing it, understanding its nature, its meaning and its raison d'être. We can then observe that to measure the value of an object requires the

---

1 The context of this section is partially inspired by [ROD 02].

necessity of establishing its intrinsic priorities, its essence and to know what it can produce: the most important is to find out its role!

In fact, in simple terms, we can say that there are two complementary ways to perceive the concept of value: a *philosophical* perspective and an *economical* perspective.

4.4.2. *Philosophical perspective of value*

The thinking of Louis Lavelle [LAV 51] in his treatise on values ties in a large part with philosophical considerations surrounding the concept of value. He explains that the word value applies wherever there is a break with indifference or equality between things, wherever one of them must be put in front of the other, in other words, considered superior and deserving preference.

In its philosophical sense, the word *value* generally has *quality* and *virtue* as synonyms. In fact, the quality of something is what defines it and makes it what it is. We could even say that a thing has no value if it is unrecognizable. Its characteristics and its role thus appear fundamental and must be recognized before the determination of its value.

In its common use, the word quality is used to define the intrinsic properties of a thing and what makes it worthy of being used. We must, however, distinguish between the quality of a thing and its virtue. According to Lavelle, the quality of a thing resides in what it is, whereas its virtue consists of what it can produce. Quality is static, and virtue is dynamic. In the second case, the value of the object is in the activity using it. This use transforms the way of evaluating the object and incorporates it to the development of the activities of man. In fact, according to this idea, we must give the object value because, even if the object is intrinsically valuable, it will remain without value if it is useless (without a role).

The comments made earlier highlight the presence of two characteristics always present today involving the philosophical problem of value: interiority and universality. Interiority determines the value in relation to the properties and qualities of objects. Universality conditions the value of objects by their capacity (virtue) of becoming useful.

4.4.3. *Economic perspective of value*

The economic idea of value is actually based on the two philosophical approaches presented earlier: interiority and universality. Because of this, two currents of thought can be linked to the economic measure of value:

– The "objectivistic–classic" current developed a theory of value that relies on the cost of production. Developed in keeping with the economic theory of Marx, this current of thought attempts to set the value of things based on the quantity of work associated to their production. The value of things is directly proportional to the time it takes to produce them, and not to their future use. The production time establishes an objective relation between objects for better comparison.

– The "subjectivist–marginalist" current maintains a functional theory. This theory strives to establish the value of things by relying on their functionality and use. The value can be determined by the capacity of things to satisfy human needs or desires. We can measure it by the intensity, time and number of people using it (the "value-functionality" concept [MIL 99]). Regardless of the work it took to produce an object, the important thing is the functionality it offers to the one using it.

4.5. The measure of value for GIS

In light of the philosophical and economical considerations briefly addressed in the previous section, we can observe the flexibility of the concept of value. In geomatics, in particular, it is possible to highlight different ways to determine the value of a GIS. These can be globally grouped into two different approaches: quantitative and qualitative.

4.5.1. *Quantitative approaches*

In the quantitative approaches, we must attempt to measure the value of a GIS mainly by responding to the following question: "Does the investment in a GIS produce enough benefits to justify the cost?" To answer this question, we must obviously be able to evaluate the benefits and calculate the costs.

4.5.1.1. *Methods of evaluation of an investment*

Models or methods of evaluating an investment are often used to measure the value of a system. There are mainly six models for evaluating an investment [LAU 01]:

– return on investment period method,

– rate of return on investment method,

– cost/benefit ratio method,

– net present value method,

– profitability index method,

– internal rate of return method.

In the following, we briefly explain these different methods. For more information, particularly on their strengths and weaknesses, please refer to [LAU 01].

The return on investment period method is the simple measure of the time necessary to reimburse the cost of the initial investment. In this context, the return period is calculated as follows:

$$\text{Number of years of return} = \frac{\text{Initial investment}}{\text{Net annual receipt}}$$

The rate of return on investment method calculates the rate of return on an investment by the adjustment of receipts produced by this investment to write it down. This method approximately calculates the accounting revenue generated by the initial investment. To calculate the rate of return, we must first calculate the average income. It is then possible to divide this income by the total initial investment to obtain the return of the investment:

$$\text{Income} = \frac{\text{Total benefits} - \text{Total cost} - \text{Amortization}}{\text{Useful life span}}$$

$$\text{Return on investment} = \frac{\text{Income}}{\text{Total initial investment}}$$

The cost/benefit ratio method calculates the return of a capital expense simply by calculating the ratio of benefits to costs. It is calculated with this simple formula:

$$\text{Cost - Benefit Ratio} = \frac{\text{Total benefits}}{\text{Total cost}}$$

Using this method, the challenge consists of being able to correctly identify the different benefits. Some of them can, in fact, be qualified as intangibles. It then becomes difficult or even artificial to make these different intangible benefits tangible, i.e. measurable in a quantitative way. Table 4.1 presents some of the commonly used tangible and intangible benefits.

TANGIBLE BENEFITS
- Increase in productivity (e.g. digital mapping) - Decrease in costs of operation (more efficient processes) - Decrease in labor (automating certain jobs) - Cost decrease in the management of geographical data - Decrease in external supplier costs - Decrease in expenses or their increase level
INTANGIBLE BENEFITS
- Better use of assets - Better control over organizational resources - Better planning of activities linked to the territory - More GI - More appropriate GI (precise, quick, exact, etc.) - Better attitude and greater employee satisfaction - Better knowledge and training for employees on GIS - Greater understanding of the territory and better decisions - Increased customer satisfaction because of GIS - Improvement of corporate image (image of innovation)

Table 4.1. *A few tangible and intangible benefits*

The net present value method calculates the current value of a payment or a flow of payments to receive in the future. This value is calculated using the following formula:

$$\text{Present value} = \text{Payment} \times \frac{1 - (1 + \text{interest})^{-n}}{\text{Interest}}$$

The profitability index is a method used to compare the profitability of different investment possibilities. This index is calculated by dividing the present value of total receipts of an investment by the initial cost of this investment:

$$\text{Profitability index} = \frac{\text{Present value of receipts}}{\text{Investment}}$$

The internal rate of return method is similar to the net present value method, but we also consider the temporal value of the money. It makes it possible to calculate the rate of return (profit) that an investment should produce.

4.5.1.2. *Other quantitative techniques*

The value of a GIS is highly linked to the economic benefits resulting from the marketing of products and services generated. In general, the total cost of production of the "first copy" is very high. It is linked to the costs of acquisition and processing of data, as well as the design and development of the system. In principle, no isolated user would be able to pay all the costs by himself. On the other hand, the marginal production cost of the "subsequent copies" is very low. This cost is generally limited to the distribution of geographical data. With this specific context in mind, other methodological appproaches of evaluation can be implemented, notably the *non-linear rates* and *decremental costs*.

The non-linear rate approach [KRE 00] suggests that among the different ways of estimating the value of products from GIS, the best way is to establish non-linear prices. According to this approach, different prices will be charged for the same type of information to different groups of buyers. The “two-part tariff” approach, for example, determines that the first part of the information sales tariff is in the form of a license, subscription, and will favor the recovery of fixed information production costs. The second part of the tariff is linked to the use of the information (number of transferred reports, bytes or layers of information reports, etc.). Krek suggest the following formula [KRE 00]:

$$P = p_0 + p_v \times q$$

In this calculation, the price of geospatial information (P) is based on fixed rates (p_0), on the price invoiced for a unit of transfer (p_v) and on the quantity of information transferred (q). However, a problem limits the realization of this method: the difficulty of correctly estimating the potential number and type of users who would be interested in using GISs and related products. It then becomes risky to calculate fixed rates for the recovery of the implementation of the geographical data diffusion system costs.

The decremental cost approach starts with the principle that, when the users search for information, they must traditionally access several sources to find what they need. This search for information is due to the scattering of information among multiple sources. Important costs are caused by this information search: movement of personnel, time, transportation, communication costs, etc.

GIS offers the possibility of finding the desired information in a centralized manner in one location. This is expressed by gains in economical, human and physical resources. In theory, the costs of locating information in a GIS are considerably lower than with traditional methods. According to this approach, the value of a GIS for users is calculated by the gap between the costs linked to the traditional method and costs for the use of the GIS, which are generally lower. The greatest gap possible between these two types of costs is then desirable.

4.5.2. *Qualitative measure approaches*

There are different criteria to determine the value of a GIS with qualitative approaches. In practice, the qualitative criteria are often simply associated with intangible benefits (see section 4.5.1.1). However, a qualitative criterion is not necessarily related to an intangible benefit. In fact, we are often tempted to transpose the qualitative measure criteria in a quantitative way. For example, if we think that the introduction of a new GIS in an organization will improve employee motivation because they can work with technically advanced tools, we may be tempted to simply measure the rate of absenteeism to determine if it is lower. We then switch from qualitative to quantitative mode. It thus becomes easier to measure financial benefits attributable to the operation.

Another problem resides in the distinction between the value of a GIS and the value of the information it returns. The GIS is, in fact, a method, a tool and, because of it, it has a value, either intrinsically (interiority) or because it is used (universality). On the other hand, this tool makes it possible to produce information, which also plays a role in organizational decision-making processes. It very often becomes difficult in practice to separate the value of the tool (the GIS) from the result (GI).

According to Milon [MIL 99], *the symbolic value of information created remains without value as long as it does not fit in the physical reality of a service*. Based on this idea, a GIS has at least the virtue of providing profitable knowledge of the land. This knowledge is liable to improve the welfare or quality of life of people. This perspective is where we think the concept of *social value of use* is interesting to qualify the value of a GIS.

The main principle of social value of use implies that the long-term welfare of a larger number of people determines the value of each good that is proposed to them. In fact, individual needs and desires are simultaneously human needs and desires in general. The needs and desires of society are the heterogenous group of individual needs and desires of all its members. The global use of an object in society determines its social value of use. Products can decrease in use and usage value simultaneously without changing in their aptitude to be used. This is the case with GIS. In fact, their social value of use can vary according to three factors:

– the intensity of the advantage that GIS can provide for the user,

– the duration of the advantage that the GIS can provide for the user,

– the number of people with needs that the GIS can provide.

On the basis of these factors, we could be tempted to attribute a higher social value of use for products fulfilling a primary need, instead of luxury products going only to a small number of people. It may be interesting to compare the concept of social value to Maslow's ideas on the classification of human needs [MAS 70]. Does the GIS then have a value as a primary need or as a luxury object? Similarly, does the GIS find its value in individual decision support, in its contribution to the global management of a corporation, or as a human and social progress development tool?

4.6. Role and value of a GIS

It is a fact that it is hard to try to understand the role and value of a GIS by separating these two concepts. They are closely linked. The GIS has a value only because of its specific role.

Now that we have defined and distinguished between these two concepts, we will attempt to compare them through the classifications previously presented. We will finally address how the role and value specifically attributed to a GIS can condition the resulting technical and organizational environment.

4.6.1. *Relational terms of the concepts of role and value*

In simple terms, we can establish that the role that a GIS can play can be rational or symbolic. In practice, the role is never clear, but we can think that there is probably predominance of one or the other role. In addition, the value of a GIS can be either quantitative or qualitative. Even in this case, the criteria of measure used can be a combination of these two measures.

It is interesting on these bases to compare the "Symbolic-Qualitative" and "Rational-Quantitative" types (Figure 4.3). Even though it is presented as a 2 × 2 matrix, it is probably best represented as a continuum, symbolized by the bidirectional arrow in the figure.

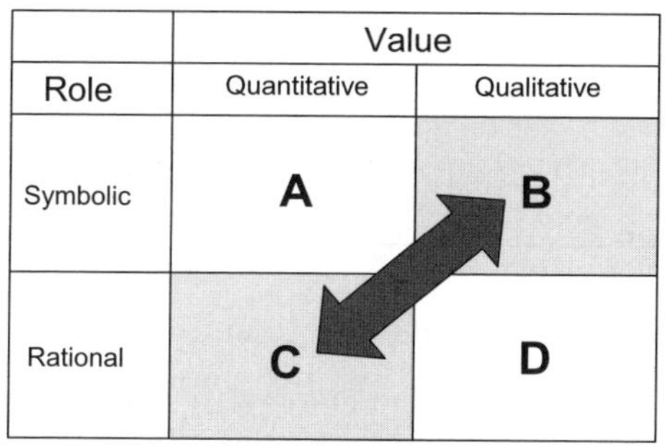

Figure 4.3. *Relational matrix of roles and values of a GIS*

The A and D situations are certainly not impossible, but nevertheless rare in practice from our observations. Situation C is probably the most common of the four types presented; it is a context where a rational role is attributed ("classic" rationality) to the GIS and for which we will have a quantitative evaluation approach. For example, the implementation of a GIS to improve the customer relationship management (CRM) of a retail chain store is a rational role for which we will be able to measure an increase in customers or a decrease in the number of complaints on an annual basis. In addition, a B-type situation, where a GIS plays a symbolic role and for which we will mostly have a qualitative evaluation, may be less frequent than Situation C. At the very least, Situation B will not be voluntarily admitted to in practice, because it is very often about strategic considerations that cannot be publicly broadcast. For example, the implementation of a web-based GIS for improving the innovative image of a city constitutes a symbolic role, which will be evaluated in a qualitative way (e.g. improvement in the relation with residents). We must remind our reader once more that all these cases are fake, and they are mixed in practice and are a difficult attempt at clarifying and put in context the concepts addressed.

4.6.2. *Possible consequences according to desired roles and anticipated value*

As we have seen, a GIS not only plays a rational role but also a symbolic role. It also often has a value that is more closely linked to operational and tactical levels than to the strategic level. In this case, there can even be a risk of losing control, i.e., to focus excessively on the GIS and forget the justification of the tool, its role in the human, organizational or social context [CAR 00].

On the basis of the notions of role and value of a GIS, we can now highlight different technical and organizational contexts directly conditioned by these parameters. We thus present three types of contexts: (1) the overadoption, (2) subadoption and (3) hyperfunctionalism.

4.6.2.1. *Overadoption*

It refers to the adoption of an innovation by a person or an organization where the experts consider that this innovation should be rejected:

> *The idea of overadoption implies that one of the roles of an agent of change is to prevent the "too big" adoption of an innovation, while attempting to accelerate the speed of the distribution process. In many cases, overadoption is a major problem [ROG 83, p. 236].*

There are several reasons that explain a context of overadoption; for example, misreading the technical innovation (GIS) for the person adopting it, the difficulty of anticipating the consequences of the acquisition or the strong attraction for a technological innovation.

It is often difficult to determine whether a person or an organization should adopt a GIS or not. According to Rogers [ROG 83], *rationality defined as the use of the most efficient methods to reach a given objective is not easy to measure*. We add that this *hierarchy of goals* may not be appropriate to determine if overadoption exists, since (a) this hierarchy can be ambiguous and (b) the acquisition of technology can respond to "other" rationalities, for example, as a symbol of prosperity and success [ROC 96].

In the case of the implementation of GIS, there probably have been instances where GISs were overadopted. Research studies have demonstrated that the *sequence* (*impulsion* and *process*) followed by a project is mainly *chaotic* or characterized by a context of "solution in search of problems" [CAR 02]. We can reasonably think that some organizations were able to acquire these solutions, without resolving any specific problem or respond to an explicit and formalized need. In addition, it is a well-known fact that some organizations with GIS implementations have subsequently let these technologies *lie idle* on desks or in storage units.

Today, we can imagine that the GIS implementation projects are managed by executives who understand the possibilities and limitations of these technologies better and who have a certain experience (positive or negative) with these technologies. The exact contribution of GISs is now better defined by decision makers and users, thus decreasing the risks of overadoption, natural and unconditional attraction for these new technologies.

4.6.2.2. *Suboptimization*

This refers to the efficient use of innovation for a part of the organization to the detriment of the rest:

> *Suboptimization is more typical of the organizational decision (than optimization). The decision maker acts on the basis of a decisional context and available information in the hierarchy of his unit or specific department. He also makes decisions locally. Such*

> *decisions cannot be optimal for the whole organization. The organization experiences the impact of all the effects, whereas a department may not be affected. Beneficial decisions of a department can create problems elsewhere in the organization that may be more important than the benefits obtained by the decision maker's department [ALE 67, p. 157].*

Suboptimization is a phenomenon that is quite often observed in organizations. Possible causes of this effect are often limited perspectives in which decision makers evolve in terms of their concerns and the information they have. In addition, reality is often simplified by managers for better comprehension.

In terms of the implementation of GIS, the perspective followed is often restricted to the technology [CAR 02]; we concentrate on *operational* activities liable to be improved by the GIS, without establishing a link with strategic considerations involving the rest of the organization.

4.6.2.3. *Hyperfunctionalism*

In simple terms, *hyperfunctionalism* can be described as mainly and excessively focused on the operation and technical constraints (e.g. computer technology) and neglecting the ambiguity and complexity of organizational, human and social aspects:

> *It is the "natural" penchant of computer specialists to forget the world and only concentrate on the single terror of the technical sanction. They design projects that will only satisfy technical constraints. [...] The technical approach of problems comes down to confuse water treatment and the pipe with the thirst and comfort of the housewife, or in other words, information processing with intelligibility, relevance and timeliness of "data". Common hyperfunctionalism that is satisfied with what works is lacking in realism to say the least [PAV 90, pp. 221-222].*

From our experience with many GIS projects [CAR 02], we observed that some of them, following the example of certain information systems development projects carried out in IT in general, were done in a *hyperfunctionalistic* spirit. In such situations, the GIS was restricted in its strictly technical role with very little integration in the business processes of the company; the criteria for establishing the value of the GIS were mainly linked to technological performance.

4.7. Conclusion

The goal of this chapter was to produce a difficult summary concerning the role and value of GIS. We hope that we have contributed to this issue. The approach followed was first to distinguish between the notions of role and value with the help of a multi-disciplinary analysis and to illustrate the close inter-relation between the concepts addressed.

The challenge with the process of this chapter comes from the fact that these notions are broken down into different fields and rooted in different paradigms. In fact, not only are the role and value of a GIS not always clear in practice, but there is not, from a theoretical point of view, a real consensus on the appropriate terms and concepts to use.

Despite the problem, it is imperative that we understand these notions better because, whether we like it or not, they are vital to a better understanding of the relevance and justification of these technologies in our lives. There is still a lot of work to be done to finalize the definition and relational terms of concepts presented in this chapter. Especially, since this task must be accomplished in the context where our organizations and corporations are constantly changing, integrating evolving GIS technologies for which new roles are probably still being created.

4.8. Bibliography

[ALE 67] M. ALEXIS, C.Z. WILSON, *Organizational Decision-Making,* Englewood Cliffs, Prentice Hall, 1967.

[BLI 89] S. BLILI, S. RIVARD, "Technologies de l'information et stratégies d'entreprise: Les prémisses d'une mutation". *Research Reports,* Administration and economy department, Université du Québec à Trois-Rivières, Trois-Rivieres, 1989, 32 p.

[BOO 71] D.J. BOORSTIN, *L'image,* Paris, Union Générale d'Édition, 1971, 436 p.

[BAI 91] V.E. BAIER, J.G. MARCH, H. SAETREN, "Décision et mise en œuvre: Une série d'ambiguïtés". in J.G. MARCH (ed.), *Décisions et Organisations,* Paris, Éditions d'Organisation, 1991, 275 p ("Les classiques E.O.").

[CAR 97] C. CARON, Cadre Descriptif des Projets D'implantation de Technologies Géomatiques dans les Organisations, PhD Thesis, Forestry and Geomatics Faculty, Laval University, Quebec, 1997, 353 p.

[CAR 00] C. CARON, "Les projets de géomatisation: contexte social et organisationnel", *Revue Internationale de Géomatique*, Vol. 10, Paris, Hermès, 2000, 167-192.

[CAR 01] C. CARON, J.P. BENDA, "Étude des projets de géomatisation au Canada", *Geomatica*, Vol. 55, Ottawa, Canadian Association of Geomatic Sciences, 2001, 309-323.

[CAR 02] C. CARON, Y. BÉDARD, "Lessons learned from case studies on the implementation of geospatial information technologies", *URISA Journal*, Urban and Regional Information Systems Association, Park Ridge, IL, USA, vol. 14, 17-36, 2002.

[CAM 97] H.J. CAMPBELL, "44-Institutional consequences of the use of GIS", in P.A. Longley, M.F. Goodchild, D.J. Maguire, D.W. Rhind (eds.), *Geographical Information Systems, Vol. 2: Management Issues and Applications,* 2nd Edn, New York, Wiley, 1997, 621-631.

[FEL 91] M.S. FELDMANN, J.G. MARCH, "L'information dans les organisations: Un signal et un symbole", in J.G. MARCH (ed.), *Décisions et Organisations*, Paris, Éditions d'Organisation, 1991, 275 p ("Les classiques E.O.").

[GUI 93] B. GUILHON, *Les Dimensions Actuelles du Phénomène Technologique,* Paris, Editions L'Harmattan, 1993, 179 p.

[KRE 00] A. KREK, F. ANDREW, "The economic value of geo information", *Geo-Informations-System - Journal for Spatial Information and Decision Making*, vol. 13, pp: 10-12, 2000.

[LAU 01] K.C. LAUDON, J.P. LAUDON, *Les Systèmes D'information de Gestion Organisationnelle et Réseaux Stratégiques*, French adaptation by L. Gingras, Éditions du Renouveau Pédagogique, 2001, 784 p.

[LAV 51] L. LAVELLE, *Traité des Valeurs,* Paris, Presses universitaires de France, 1951.

[MAS 70] A.H. MASLOW, *Motivation and Personality*, New York, Harper & Row, 1970.

[MAR 76] J.G. MARCH, The Technology of Foolishness, in J.G. March, J.P. Olsen (eds.), *Ambiguity and Choice in Organizations*, Universitetsforlaget, Oslo, 1976, 408 p.

[MAR 88] MARCHAND J. *Le jeu de la réussite*. Les éditions Le Griffon d'argile, Sainte-Foy (Québec), 117 p. (Collection Philosophie) 1988.

[MAR 91] A.C. MARTINET, "Préface", in J.G. March (ed.), *Décisions et Organisations,* Paris, Éditions d'Organisation, 1991, 275 p ("Les classiques E.O.").

[MIL 99] A. MILON, *La Valeur de L'information: Entre dette et don,* Critique of information economy, Paris, Presses Universitaires, 1999.

[PAV 90] F. PAVÉ, "Peut-on rendre certaines certaines incertitudes? Ou: De l'art de guider le désordre", Social evaluation of new information communication technologies, in J. Berleur (ed.), *Actes des 4e Journées de Réflexion sur l'Informatique,* Namur, Presses Universitaires de Namur, Nov 30–Dec 1, 1990, 267 p.

[PIC 90] F. PICHAULT, *Le Conflit Informatique,* De Boeck-Wesmael, Bruxelles, 1990, 259 p.

[ROC 96] S. ROCHE, C. CARON, Y. BÉDARD, "Vers une approche plus complète du rôle de la géomatique dans les organisations", *Revue Internationale de Géomatique*, Vol. 6, Paris, Hermes, 1996, 73-92.

[ROD 02] O. RODRÍGUEZ-PABÓN, S. ROCHE, F. BROCHU, C. CARON, J.J. CHEVALLIER, "Établissement de la valeur d'une infrastructure d'information géospatiale selon la perspective de l'interactionnisme social", *Géomatique 2002*, Combined international symposium "La théorie, le traitement et les applications des données géospatiales", Ottawa, July 9–12, 2002, 7 p.

[ROG 83] E.M. ROGERS, *Diffusion of Innovation,* 3rd Edn, New York, Free Press, 1983, 453 p.

[SCA 89] V. SCARDIGLI, "Nouvelles technologies: L'imaginaire du progrès", in A. Gras, S.L. Poirot-Delpech (ed.), *L'imaginaire des Techniques de Pointe: Au doigt et à l'oeil,* Paris, Edition L'Harmattan, 1989, 229 p.

[SON 99] D. SONNEN, H. MORRIS, J. SWEENEY, *1999 Worldwide Spatial Information Management Markets and Trends*, International Data Corporation (IDC) Report #W21065, 1999.

[VAN 01] J. VANDERMEER, "What's the difference between M-Commerce and L-Commerce?", *Business Geographics*, 2001, available at: version www.geoplace.com.

Chapter 5

Social Use and Adoption Models of GIS

5.1. Introduction

The problem discussed in Chapters 3 and 4 shows how geographic information technologies (GITs) or geomatic technologies offer great potential to improve corporate performance. However, to use them, we must adopt them, which inevitably implies a will and organizational strategy to change the dynamics of the social and professional organization, sometimes drastically. Research becomes essential to better define and understand the processes of GIS adoption and organizational change, and thus support these change processes (this topic is addressed in Chapter 6). Ideally, it must also subsequently lead to the development of tools to assist decision makers in the determination of what is relevant for their organization.

The main objective of this chapter is to provide a state of the art on GIT diffusion and adoption problem. An important part of the comments goes beyond the strict context of GI and generally involves "IT", "communication tools" or "interactive technologies", presented in the form of technical objects occupying our organizations. The

---

Chapter written by Stéphane ROCHE and Benoît RAVELEAU.

different diffusion and adoption models of IT, often put in prospect because of the inherent characteristics of GI, are analyzed, in particular. More specifically, this chapter analyzes the (organizational, social, cultural, spatial, etc.) factors influencing the diffusion and adoption of IT, in general, and GIT, in particular.

We focus mostly on the social aspects of these interactive communication technologies in a context where technical mediation is crucial in daily activities. With the help of a questionnaire focused on the *use* of technologies, which often seems to be obvious, we will study the question of their *social appropriation* and examine *the relation of individuals with technical objects*, and their content, in the light of different theoretical and methodological approaches. Our objective is less about defending a specific thesis than it is about reporting on the different theories and methodologies used in research. To do this, we will use the work from Pierre Chambat published in 1994 titled "*Usages des TIC: évolution des problématiques*" (ICT Use: Evolution of the Problems) [CHA 94a,b], a paper that we will use as the framework for our synthesis. This paper is interesting because it goes into the different approaches of usage sociology[1], through a dialog between them. We will go back to these distinctions by quickly presenting the diffusion and innovation approaches, and then concentrate more on the appropriation approach. In fact, the latter seems the least well documented of all, even though it groups a large quantity of studies and research projects.

5.2. From ICT to GIT

We start this chapter by directly discussing the GIT world. It seems absolutely necessary for us to start with a quick overview of what GIT has in common with IT to ensure a transition with the previous chapters, as well as their specific characteristics involving this diffusion and adoption problem. In fact, even though GITs can be

1 Our definition of "sociology of use" is not a sub-discipline or a research trend per se. It mostly represents a series of research studies with common concerns, fitting in the field of social uses of media and technologies.

considered as somewhat of a subset or a specific listing of information and communication technologies (ICTs) and, consequently, have a certain number of theories and concepts developed for ICTs, they nevertheless have inherent characteristics deserving mention and consideration in the analysis. The objective here is to highlight these characteristics, while showing that we are justified in applying the models and theories developed for ICTs to GITs.

5.2.1. *GIT: a subset of ICT*

GITs are very often, and justifiably so, considered as a specific type of ICTs. We maintain the idea that they are a specific subset of IT, in reality. This affiliation is explained and justified at two levels.At the first level, GITs can be considered as IT by *nature*. GITs constitute the point of convergence between spatially referenced data acquisition techniques (related to geodesic science disciplines), the methods and models of spatial analysis and graphical and cartographical representation (related to geography) and IT (especially from the evolutions that the IT field has experienced in the last two decades). Therefore, GITs are based on ICT's technological infrastructures (networks, databases, information systems, etc.). In addition, current technological evolutions (e.g. Internet, mobile and wireless technologies), and especially the phenomena of technological convergence, keep bringing them closer to IT while potentially widening their range of use and applications (this is demonstrated in Chapter 3, in particular). At another level, because of their *destination* and linked to the nature and methods and purpose of their uses, GITs must be considered as IT. Their main objective is to inform and to communicate. They are the first and foremost technologies designed to acquire, maintain, update, represent and diffuse information toward a varied public. In this, they are similar to IT in general.

5.2.2. *Characteristics of GIT compared with ICT*

GITs have inherent characteristics that we must consider in analyzing their diffusion and adoption methods. These characteristics are not only limited to the "technical" nature of the GI as we were always led to believe. The question of location, georeferencing and the geometric structure of graphical (or geographical) objects is important and characteristic of the field of GI. In fact, there are organizations and representation models specific to GITs, research and analysis tools adapted to GI, for example, and they are examined in the other ISTE publications. We are not saying that these more technical characteristics have no effect on the methods or development of GIT adoption processes. However, we want to emphasize the spatial dimension of GITs in a social approach rather than a technical one. The role of spatial dimension in the analysis of uses and methods of diffusion and adoption of GITs can be explained on three levels: the social construction of technologies, information production and uses.

The development of GITs is similar to a social construction that is deeply rooted in an organization, a society located in a given territorial context [HAR 97]. Because of this, all the territory (its organization, characteristics, the relations of social groups with their sphere of reference, etc.) constitutes decisive variables for the diffusion and adoption processes of GITs [CAM 96, ROC 00]. The spatial dimension is expressed by physical anchoring of diffusion and adoption processes in a given territory and by cognitive representations somehow characterizing the relations between the people involved in these processes, and the support space for their activities [CAR 01]. However, GITs produce spatial representations, territory models; they are, in a way, the reflection of the relations between the men who design them and the territories on which they have to reflect, to make decisions and that they must provide for. At the same time, GITs have the capacity to modify these relations. Prior research projects have shown, for example, how confrontation between cognitive spatial representation from GIT users and physical spatial representations proposed through technologies could contribute or limit the adoption of GITs depending on whether there is compatibility or incompatibility between them [BAC 02, ROC 00]. Finally, jobs and fields of use of GITs are specific and contribute in separating them from ICT or, at least, put them in a category by themselves. First, from a professional standpoint, the use of GIT affects a wide variety of territory businesses (see Chapter 2). As for personal use, a wide range of people use GIT in generally simplified forms today: web mapping, GPS navigation embedded in vehicles or through PDA such as Palm, position determination technology with the help of cell phones, etc. In all cases, the fundamental element remains space and GITs, in this instance, are the new keys to see and understand, along with the obligatory interfaces.

5.3. For a social and technical approach dedicated to practices and uses

The multiplication and complexification of ICTs should not fool anyone; use and appropriation give its value to technology. The main factor of their efficiency and return on investment resides in the way the stakeholders individually and collectively use them for their domestic or professional uses. The contemporary renewal of social thinking on social and technical systems supposes that we carefully reflect on what people do with the instruments at their disposal. Such is the ambition of sociology of uses and technology in which we fit. In this case, the focus is not on the technological offering, although quite fascinating, but on the concrete, regular use from users in their domestic or professional worlds. This comprehensive methodology is inherent to the web-based sociological tradition, in which a social and technical system can analyze and explain itself by taking into consideration autonomous stakeholders, concrete situations made up of human connections, of the meaning created by stakeholders and rules of the game that hold and evolve the given set. Before proceeding further, several notions should be discussed.

5.3.1. *The notion of use*

The notion of use is ambiguous. It recognizes different explanations, implying completely different conceptions from technical and social approaches. To clarify the term, Jouët [JOU 93b] distinguishes the notions between the use and practice: *the use is more restrictive and goes back to the simple usage whereas practice is a more elaborate notion covering the use of techniques as well as behaviors, attitudes and representations of individuals directly or indirectly relating to the tool.* Very often, this differentiation is not accomplished, and these two terms are confused. On the other hand, the "social uses" expression seems to be establishing itself. Lacroix [LAC 94] proposes the following definition: *social uses are modes of usage occurring with enough recurrence and in the form of habits sufficiently integrated in every day life to fit in and possibly resist in terms of practices specific to other concurrent or related practices.* Other researchers from various disciplines have been interested in sociology of uses. Some have specifically analyzed the relation between offer and demand by showing how the development of uses happened at the junction of the technical and social approaches. That is why in this chapter, we assert with Millerand [MIL 98] that *usage goes back to the use of a media or a technology that can be detected and analyzed through specific practices and representations; usage becomes social when it becomes possible to capture social conditions of emergence – because they are stabilized – and in return establish the methods by which it participates in subjects' social identities.*

5.3.2. *From technical to social commonality*

In research involving social uses of technologies, we can broadly consider that the debate comes down to tension between supporters of technological determinism and supporters of social determinism, and more subtle positions characterized by most researchers in sociology of uses. In fact, as noted by Scardigli [SCA 94], more than 30 years, research has gone from questioning centered on technique to uses. The 1970s, when priority was economic growth, were marked by *the technical progression of all fields of industrial and social activity.* The research carried out in that period came from a "techno-logic of social change", according to Scardigli. The two dominating explanatory models of the time are technicist (or voluntarist) and evolutionary models, which suppose that new techniques inevitably transform society in the direction planned by their designers. The expressions "ICT impact" or "ITC effects", frequent in scientific literature at the time, clearly reveal this sort of primacy of the technical over the social inherent to technological determinism. We have seen in Chapter 1 that this position is not completely absent from current research involving relations between GIS and society. The prophetic nature of the speech that accompanies ICT diffusion is also due to the problem with understanding the related uses. This view considers that the implementation of new information and communication information techniques will "push" social organization, whereas reality shows the complete opposite. The technical tool fits in a context of accompanying transformations of practices and uses. In the 1980s and 1990s, it became obvious that the linear diagram at the basis of technological determinism cannot account for actual usage.

Technological determinism was then gradually abandoned for much more nuanced positions. Researchers stick to *the purpose that each social micro-stakeholder will give his life* study [CHA 94a,b] and refusal, misappropriation or bypass behaviors of uses imposed are questioned. Scardigli [SCA 94] speaks of a "socio-logic of social change" to characterize this trend in terms of research. He also mentions Touraine [TOU 92] and the "rediscovery of the subject" to understand this transition *products – goods or services integrating new technology – to the person, simultaneously consumer, worker and resident of a city*. Ignoring the technical constraints would lead to a simplistic symmetric position of social determinism, but in reality, the positions are not as distorted. Sociologists often emphasize the fact that media practices and the uses for these communication tools are inseparable from daily activities: *communication practices, contrary to leisure practices often perceived as residential or compensatory in relation to alienating work appear as central, because ICTs are diffused in all daily activities* [CHA 94a,b].

According to the different approaches of sociology of uses, every day life, or the context of daily domestic or professional life in which technological innovations diffuse, appropriate or institute, is perceived and considered differently. In fact, according to the analysis models, daily life is mainly considered in terms of standard of living (the important socio-demographic is revenue), way of living or lifestyle (the personal stories of the subjects and their living conditions in terms of revenue are integrated in the analysis). This reflection on daily life invites us to redefine the study of uses in a wider reflection on the evolution of ways of living and lifestyles in social, economical or even political terms in the case of GIT, and probably also in territorial terms. Chambat [CHA 94a,b] invites us to take the plunge between *micro-social behaviors and macro-social trends*. Proulx supports setting the media reception study in social, economical and production policies for media content conditions and invites the introduction in the analysis of the media globalization context. The daily life study implies expanding the analysis to the consideration of deep social trends and greater disruptions in global society contributing to the evolution of lifestyles.

Some researchers in the field of GI have, in fact, been part of this debate. Campbell [CAM 97], for example, clearly emphasized that practitioner methods, and research projects developed to understand the diffusion process of geomatic technologies have gradually fallen into three different paradigms: technological determinism, managerial rationalism and social interactionism (see Chapter 4).

5.3.3. *Three levels of understanding social uses*

ICTs or digital technologies generally include all objects represented by these technologies. And yet, when she examines research involving the social uses of the technologies more closely, Millerand [MIL 98] observes that at least three different conceptions of the notion of use can be distinguished: as "tools", a "social sign" or a "device":

– the "tools" object: research projects developed here lead to a conception of use as a generally functional and powerful use;

– the "social sign" object: research projects considering them as "social signs" define use as the generally distinctive expression of the social status;

– the "device" object: substituting the notion of object with the notion of "device" represents a conception of use as a more or less accentuated subjection to social standards [CHA 94a,b].

Even though the common point is that these tools, services and devices were created for maximum commercial profitability, this distinction presumes the existence of different types of relations to objects for users. Even the objects that can be qualified as "common" still carry a relation to the object. The example of the cell phone is particularly interesting and shows how one object can serve various designs such as instrumental (through professional or personal appointments), recreational (games) or social (SMS, family or friends can constantly contact you). The cell phone is also used to represent an obvious "social distinction", even though today this object of communication has become a daily partner, a crutch, that accompanies the user in his every day life. For other objects, such as television or the personal computer, the question concerning the co-existence between "object file" and "usage file" can be raised. Pronovost [PRO 94] questions the possibility of integrating uses instrumental to a technology apparently dedicated to entertainment (e.g. television). Recently introduced technological innovations make this problem even more complex because they emphasize the gap between the tools and the services they offer. This is the case with several technologies emerging over the last 10 years in homes: Internet, webcam, interactive television or wap telephone. As opposed to previous home devices, these technologies offer a diversified range of uses, in turn game buddy, babysitter, decoration, etc. Their use may aim for the illusion of a presence or enabling it, by offering the possibility of communicating with other people located elsewhere. Clearly, this development of technological objects fits into the larger context of increase in the services sector. The question is then raised of a possible reappraisal of analyses centered on equipment, for the benefit of analyses centered on services, commutation and social connection.

The questions raised by the national and international geomatic community on the nature and forms of use of GIT do not escape this phenomenon. From questions typically centered on the development of technological tools, a shift has gradually occurred toward the implementation of information systems, actual devices for ensuring the life cycle of GI. The concept of GI infrastructure has developed at all scalar levels in the last 10 years. These GIIs, as they are sometimes called, are materialized by the implementation of policies (access to data, copyrights, standardization, price setting, exchange protocols, etc.) where organizations must interact with technological infrastructures to control, in a more efficient manner, production, management and especially the use of products and services from GI [ROD 02]. Permanent efforts are therefore carried out to attract an increasing number of users to GI (embedded navigation, web mapping, mobile GIS, etc.). Studies today involve services coming from GI and their insertion in practices and social relations more so than the technical tools themselves. The case of the GPS is interesting because it is typically a geomatic technology with many uses. GPS receivers were first developed for exclusive military use, and they are today on the verge of becoming mass consumption technologies for uses, in places and by people with very diversified profiles: road and nautical navigation, outdoor sports, fishing, hunting, as well as goods and services location (GSL), etc.

5.4. Theoretical approaches of diffusion and innovation

The evolution of theoretical approaches and methodologies concerning the question of geomatic technology uses is characterized by a conceptual movement similar to the one we find in mass media sociology: from the analysis of effects to the analysis of the reception. At first, the studies focused on the analysis of diffusion of technologies in terms of "adoption" and "acceptability" by users, which starts with the anteriority of the technique and is based on a linear diagram. Research then set out to analyze technical innovations as social assertions, and to study their appropriation from user point of view, in other words analyze the formation of social uses through their implementation[2]. We observe a shift from the interest centered on

2 This approach, relatively new, groups a large number of empirical research projects, emerging from the context of the profusion of media and technological innovations

technology toward users, accompanied by an increasing use of ethnographical or microsociological methodologies. These approaches of sociology of uses study the process and structure implemented with the expansion of ICT and social logic occurring in the field of communication. Economic characteristics combine with these technical criteria as these new technologies are developed by industries in the communication sector supporting trade logic. In this brief presentation, we will limit ourselves to three approaches of sociology of uses: the *diffusion*, *innovation* and *appropriation*. Because of its importance in contemporary debates, we will dedicate a separate section to appropriation.

5.4.1. *The diffusion approach*

Research projects developing this type of approach do not focus on the product design phase but on the analysis of technological innovation adoption during this diffusion. Questions on research involve how innovations are diffused, as well as who are the adopters, and what the impact of their adoption through changes in practices is. This diffusion approach emerged from the theory of innovation diffusion from Rogers [ROG 83], first published in 1962. Early on, this author instilled a certain number of research projects in several sectors of activity [KAT 71]. These works by Rogers fit in a long anthropological tradition known as "diffusionism". The main instigator is Kroeber [KRO 23], an anthropologist mentioned in [BEL 92], who focused on the penetration of technical innovations in the cultural fabric.

In a study by Rogers [ROG 83], diffusion is perceived as a five-step process, from when the user first gets exposed to innovation to the confirmation or rejection of the adoption:

– knowledge (the person is exposed to innovation and acquires some notions of its operation);

(tape recorders, cell phones, videotex, interactive television, etc.), characterized by broken up theoretical and methodological positions.

– persuasion (the person initiates a position about innovation);

– decision (the person tackles activities that enable him to adopt or reject the innovation);

– implementation (the person uses innovation on a daily basis and evaluates it);

– confirmation (the person tries to gather information enabling him to confirm his choice).

These are the characteristics of innovation as they are perceived by the people who determine its rate of adoption. Five attributes characterize an innovation: its relative advantage, compatibility with the values of the affiliation group, complexity, possibility of testing it as well as its visibility. The users are classified into five typical profiles: innovators, first users, first majority, second majority and latecomers. Data collection and analysis techniques used in this approach are mainly quantitative sociological methods and are presented in the form of surveys using questionnaires. In short, *in addition to generating numerous empirical research projects, the major advantage to the diffusionistic model is the capability of describing the complete social network of the circulation of an innovation within a society* [FLI 95a,b]. As an illustration, we can refer to the research by Caron *et al.* [CAR 85] on the diffusion and adoption of the domestic micro-computer in Quebec. This study reveals the importance of inter-personal contacts in the computer adoption decision, notably by emphasizing the role of the workplace and home in creating the interest for the adopter.

Nevertheless, this model was the subject of much criticism, especially for its pro-innovative character. The typology of "ideal type" adopters prevents us from considering abandonment, once adoption has been decided. The main criticism with this approach involves the status of the technique. According to Boullier [BOU 89], Rogers contributed spreading a false conception of the notion of diffusion that the diffusion of an innovation would only occur when the innovation was completed and ready to be adopted. This vision reveals indifference in users accepting or rejecting the innovation.

These studies gradually integrated the notion of "reinvention" to take into consideration the way in which users modify the device they adopt. In terms of understanding users, the lack of socio-demographic variables that usually retained to describe innovators was greatly emphasized.

In research projects involving GIT diffusion, the Rogers' model was used in the early 1990s in the United States to try and understand these diffusion processes as well as their determining factors in organizations [MAS 93, MAS 95]. Rogers himself was involved with the first discussions specifically dedicated to GI [ROG 93]. A second series of case studies developed in Europe toward the end of the 1990s enabled us to expand the limits of the Rogers' model and to adapt it to the specific case of GI. The role of geographical variables and the GIT diffusion and uses, for example, was clearly highlighted [CAM 97, ROC 99a].

5.4.2. *The technical innovation approach*

Research studies grouped under the name of sociology of innovation adhere to the technical innovation processes study, i.e., the conception phase of products involving decisions and technical, social, economical and political choices. In France, this trend is mainly coordinated by sociologists (Callon, Latour and Akrich among the most well known) from the Centre de sociologie de l'innovation (CSI, Center of Innovation Sociology) of the Ecole des Mines de Paris. Their research studies are a part of the school of translation that will be widely discussed in the next section. For most, these studies rely on case studies on technical innovations that did not get implemented [CAL 86]. To clarify, we offer two examples of studies dedicated to the implementation of groupware and office automation tools in the media.

The first one involves the implementation of Lotus Notes. This groupware was designed as a facilitator for the cooperation of stakeholders at work, over traditional tools of individual productivity or operations processing. Executives gave a very positive overview of how this technology was going to modify the organization and work conditions. The decision to go ahead seemed well founded. The project was relevant, the investment was measured, the technology was matured and the strategy was efficient. However, in practice, Orlikowski observes that setbacks were major, because management did not consider the key factor of success for technologies in organizations: the concrete way in which we use it to work. Practice is the deciding factor on how the value, meaning created by the people, efficiency and technology are captured.

The second example involves ICT diffusion within a regional daily press for the purpose of helping in the integration of input and layout functions [SAI 99]. This evolution is often supported by executives. Sometimes, however, beyond this trend, professional logic must be considered. Journalists resist using IT. Any new intrusion from the computer and its software brings back to questioning what the origin of their job is: investigating and researching the event. They have a diminishing vision of the computer, which locks them in their writing. They are not looking to develop new knowledge connected to the use of computers. They will have limited use of computer because, with computers, not only their sedentation is put into question but also the nature of the job itself. On the other hand, workers in book industry, who have feared impending doom in the industry, will actively negotiate strategic positioning with the introduction of new technologies. They will demand to remain the guardians of a professional skill, which is now becoming a skill on the new IT tool. And after a long training phase for the computer system and layout software, they will continue to develop new knowledge individually and collectively.

From this perspective, technology has no meaning, value or consequences by itself: the individual and collective practice will be the deciding factor. An analysis of uses and forms of adoption reveals that the social construction of category-specific interests, as well as the concern for preserving the trade culture, is resisting technical evolution. By itself, it cannot transform the trade organization of an industry. To leave out this rule of social and technical systems in the workplace is to make three simplistic and illusory assumptions: if the stakeholders have the technology, they will use it willingly (through fascination or recreation) or reluctantly (through tasks); they will use it as it was intended by the designers; this compliant use will produce the initially expected results. The basic experience of a user leads us to think that we spend most of our time taking on this tool and the object asking ourselves how we can best use it without wasting too much time. None of us passively obey injunctions from machines and tools surrounding us or specifications imposed by developers. If we must use it, we ask ourselves where, when, with whom and how without losing ourselves in pre-existing procedures.

Studies by Vedel and Vitalis [VED 94, VIT 94] are a part of the same sphere of influence, even though they mostly develop a social policy of uses appealing for a re-integration, during design, of the face of the citizen in the user model. Their analysis raises questions on the institutional representation of the user in the technical innovation process and of the role of public institutions in the choices concerning the technical offer. This approach is interesting in the fact that it attempts to link the macro-social dimension of the analysis of the technical offer with the micro-social dimension of the actual study of use. In this context, this approach attempts to create a link between the methods exclusively centered on the analysis of innovation processes and those based exclusively on the analysis of uses in reality, and succeeds in painting an interesting picture of the different "logics" presiding over forming of uses. We will come back later to the social policy of uses in the appropriation approach.

5.4.3. *The school of translation*

In the sociology of uses, the translation model is the current dominating trend. It analyzes the use of new technologies from the translation occurring between the person and the technical object. The object is to identify the interactions between stakeholders involved in the development of the innovation. The operating instructions play a switching role from the machine to the user. The user ignores certain elements of the device, interprets others or submits to the social and technical style of the machine. The translation is done through permanent negotiation between the tool and the user. Technical objects are defined as social assertions resulting from interactions between the different stakeholders or groups. In this perspective, *a technical innovation process is presented [...] as a competition between different projects and ends when one of them becomes predominant* [VED 94]. The notion of mediation is central in the translation approach; it demonstrates the technique and social muddle, notably through the different representations of the user in the technical device.

The innovation process is considered as a succession of ordeals and transformations where a series of stakeholders are related. The review of discussions that come with any innovation process has shown the close overlapping of technical and social contents, which explains that two competing projects generally involve a confrontation of two conceptions of the world and society. The idea is that one social and technical system will stabilize once a series of translation, enrolment or profit sharing operations leads to the constitution of alliances and/or oppositions between different stakeholders. *The job of the sociologist then consists of describing the operations by which the starting scenario, mainly presented in a discursive form, will gradually, through a series of translation operations transforming itself, be appropriated, carried by an increasing number of entities, human stakeholders or technical devices* [AKR 93a].

In other words, the study of the consecutive modifications of a statement (or a program of action) will lead to a final object. These transformation operations are analyzed in terms of "in-scription" and "de-scription" of the environment in innovation (i.e., the technical device). We remember the work of Aitken [AIT 93a] who, in his story of radio, demonstrated the specific role of translator played by the different stakeholders (Hertz, Lodge and Marconi), notably by updating translation operations occurring in the different skill sets of each. Technical tests (e.g. usability tests), associations with other stakeholders, experimentations with users can be interpreted as *confrontations between the hardware environment and the environment described by the move* [of the device] [AIT 93a]. The stakeholders, natural elements and technical devices come out transformed from these confrontations. That is how innovations lead to knowledge, technical devices and forms of organization through technical mediation.

It is mainly because it is well adapted to the analysis and understanding of processes in which numerous stakeholders with different profiles are involved that the translation model was used by some researchers in the GI field. Studies from Latour [LAT 89], for example, were mobilized in reflections dedicated to the process analysis of GIS social construction, demonstrating the interest of using the translation approach and the theory of networks of stakeholders to understand the processes in the field of GI [HAR 97, HAR 98, ROC 99b].

One of the main criticisms for this approach involves the low priority it gives the role of practices, i.e., the user's action on how the technical object is shaped. Flichy attempts to exceed this limit [FLI 95a,b]. He defines a social and technical context of reference, a "context of operation" going back to the object's functions and to the technical use and a context of use, referring to social use. The fusion or combination of these two contexts leads to a new "social and technical context" once the innovation is stabilized. Finalizing the context of operation involves not only innovators but also several stakeholders, and the context of use is not set; it can be transformed by the first uses and periods.

5.5. The appropriation approach

This approach is distinguished from the other approaches on many levels. Contrary to the technical innovation approach focused on the moment of conception of technical objects, the appropriation approach puts its analyses at the level of implementation in social life. In addition, contrary to the diffusion approach linked to the study of technology diffusion process through the evolution of a rate of adoption, the study of uses in terms of social appropriation of technologies goes back to the analysis of their development from the "point of view" of users. Research will demonstrate that people appropriate these tools for personal development, work accomplishment or sociability. Case studies done in the late 1990s within municipal organizations, for example, highlight some of the underlying motivations with the appropriation of GIT by stakeholders involved in land management: affirmation of the role and status within the organization, distinction from competing stakeholders, etc.

5.5.1. *Antagonisms between espousing technology and usage technology*

Contrary to the diffusion approach of innovations focused on observing and explaining the disparities by differentiating user profiles, the appropriation approach highlights the disparity of uses and users by showing the social construction of usage, notably with what it means for the user. The question of the object status comes down to understanding what it represents for its user(s), how it fits in a specific environment and among the existing practices in the context of daily life, inseparable from social trends involved in constructing lifestyles.

In their studies, sociologists Argyris (Professor at Harvard University) and Schôn (Professor at Massachusetts Institute of Technology) carefully distinguish *theory espoused and theory in use* – what we profess to be our way of choosing and acting – and theories of use (*theory in use*) – what our practices, actual and in practice reveal about our behavior and our representation [ARG 78, ARG 95]. What we buy and install is set and defined in advance (commitment or espoused technology); integrated modules in hardware and software include support and procedure manuals. What we use depends on the situation, the location, the moment (technology of use); the specific and delimited characteristics that we use according to our needs, skills, attention and interests. We cannot mix or manage the same way, or measure the first one as if it was the second one.

The two stakeholders discussed earlier demonstrated that there is often contradiction between the two, hence conflict and possible appearance of chronic inefficiency with users. This separation between both levels (in theory and in practice) weakens the collective learning in the social organization. These two organizational analyses observe that practitioners are rarely aware of this shift between what they say and do with their technology and what their actions say about their behavior. The proposition from Argyris and Schôn is clear; what matters in the evaluation of the global efficiency of a technology is not the commitment to the principle but the technology of regular use in the real world.

5.5.2. *Appropriation or modes of usage structure*

A certain number of studies in information and communication sciences propose an approach in terms of appropriation, in other words, considering the ways in which the stakeholders appropriate technologies and make them usable. A first element is technical knowledge and command. However, if such a command is a necessary prerequisite for use, it is not enough to explain uses. In fact, the observation of practices in the field shows that this acquisition alone is not sufficient or rather that the acquisition of new cognitive skills to command ICTs should refer to the context, which is a structuring element of ulterior practices and knowledge.

The numerous theoretical and empirical research studies carried out for this approach center on various interests and research questions. One of the main research questions raised in this wide field of research consists of analyzing how different uses are made according to social groups, notably through the review of "usage meanings". Studies include questioning centered on the role of prior practices (usage file, generational phenomenon), identity construction phenomena as well as technique socialization. The different studies have shown the vital role of representations in the structure of practices, some in particular have set out to demonstrate the influence of the technical imaginary, and revealed the range of the symbolic weight of these new communication tools.

The problem with lifestyles crosses the studies involving the way ICTs restructure relationships between private and public space, work space and outside of work, leisure and domestic life. On a larger scale, the role of technologies on the technical progress of communication practices and its incidences on a cognitive level are the subject of problems exceeding the appropriation analysis of a tool by a user, to a questioning on long-term consequences of this new form of communication, increasingly mediated by the technical. Questions of technical mediation and social connection are addressed in several works. Finally, questions about the different "faces" of users, particularly the face of the "active user" brought on by the views surrounding "new technologies", are the subject of important differences of opinion, contributing to relaunch the debates on the differences between old and new media.

In a professional context, these questions illustrate the relationship between the technological offer, organizational strategy and social appropriation of tools. One of the ways to understand the integration of a technical system is to take an interest in the appropriation of ICTs, i.e. to the different modes of usage structure. The appropriation marks an irreversible threshold in the structure of uses. It occurs only after the discovery of technology and the structure of its use phases. The appropriation is created as the object is developed for the user. What starts out as a lucky encounter will gradually find synergies with other daily practices. The appropriation is created in a centripetal way from the subject [BER 82]. The context in which usage is developed influences appropriation time. Studies involving usage structure of information technology in residential, public or professional contexts demonstrates a multitude of types of discovery and structure.

We understand that ICT usage structure is not limited to the one planned by the corporation. If we listen carefully to what users are saying, we also understand that this structure work that includes the mental acceptance of the tool is also an individual "job" (in the strongest sense of the word) apart from the professional environment. We can learn alone, collectively, at the office or remotely and each situation has different meanings. Challenges with adaptation are very different depending on the employees. Whether they see the computer as a promoting element or devaluation of their job or whether they are in an uphill or downhill period.

In a prior study, based on the analysis of different GIS projects in cities in France and Canada, Roche [ROC 97, ROC 98, ROC 00] highlighted the organizational and social challenges linked to the social appropriation processes of GITs by the different direct or indirect users. The case studies explained the variables that influence these appropriation methods, and clearly showed how real GIT uses in the real world are not always on the same wavelength with the objectives assigned by GIS carriers and theoretical views, and the range and usefulness of these technologies. These studies also show the influence of the symbolic weight in the way that stakeholders appropriate GIT, bring them into their professional logic and create very different methods of use than the official views [ROC 96], (specifically shown in Chapter 4).

5.5.3. *Determining variables of appropriation*

Several research studies were carried out in the field of IT, organizational sociology or management, in particular, for analyzing and understanding the models of technology adoption and the variables influencing these models. A rich synthesis is proposed in a recent article [LEG 03]. These models have in common the consideration of cognitive variables and social and cultural characteristics of the context as central elements in adoption processes. In other words, a typology comes out of these studies, organized in different categories depending on what they are linked to:

– user profile,

– the technology itself,

– the context of use (Chapter 10 illustrates this point in the case of PPGIS),

– user perceptions,

– the nature and form of information diffusion.

For reasons mentioned at the beginning of this chapter, appropriation processes of GITs are restricted by variables of the same nature as those identified earlier and generally determine the appropriation processes of ITs. A few prior studies made it possible to highlight the specifications of geomatic technologies with regard to this appropriation question [CAR 01, ROC 00]. Spatial dimension has revealed to be one of the key points. This spatial dimension is not only due to the spatialized, localized or even georeferenced character of the information processed. This character, nevertheless, generates variables linked to the nature and form of diffused information. It is expressed on a cognitive level through the role that the differences between cognitive representations and physical spatial representations proposed by GITs play in the adoption process of GITs (which goes back to the variables linked to the profile and perceptions of users). However, it is also expressed by the influence that the geographical context plays on the content and form of a local geographical database (geographic coverage, spatial and/or thematic completeness, etc.). Here, the context in the widest and deepest sense of the word is the determining factor.

	The diffusion approach	**The innovation approach (including the school of translation)**	**The appropriation approach**
Preferred objective of the approach	Observe and explain the disparities in terms of rate of equipment and practice.	Highlights the social dimension of the technical innovation process. Concentrate on the moment of conception of technical objects.	Analyze the structure of uses from the user's "point of view". Study the implementation of technical objects in social life.
Main representatives	Everett M. Rogers, Bardini I.G.: Onsrud, Masser, Campbell	Bloor, Sociologists from CSI of the École des Mines de Paris. (Callon, Latour and Akrich), Ecole de la traduction, Flichy, Vedel and Vitalis I.G.: Harveys, Chrisman	Chambat, Argyris and Schôn, Perriault, Proulx, de Certeau, Mallein and Toussaint, etc. I.G.: Roche, Campbell

Preferred methods of investigation	Sociological quantitative (investigation) methods and correlation calculation. Study of the diffusion process of technologies through the evolution of a rate of adoption.	Case studies of technical innovations. Ethnographic techniques for "the follow up of the innovator at work". Analysis of views to discern the different transformations of statements.	Especially the qualitative methods (participating observation, in depth interviews, etc.). The quantitative sociological methods as well as the analysis of views are also used.
Main interests or advantages	Numerous empirical research projects. For a description of the complete social innovation circulation network within a society.	Shows the social dimension of technical innovation. Identifies interactions from the different stakeholders contributing to the development of the innovation.	Shows the social construction of use, notably through its meaning for the user. Explains how different uses in relation to social groups are constituted.
	Highlights the role of interpersonal contacts in the decision and the insufficiency of socio-demographic variables to describe innovators.		
Limits or criticisms	In this model, the diffusion of an innovation only occurs when the innovation is completed and ready to be adopted.	No consideration for the role of practices, i.e., the action of the user on how the technical object is formed.	Rupture of research objects and problems. Important variations at theoretical and methodological level.
Keywords	Diffusion, process, steps.	Innovation, mediation, translation, social and technical system.	Appropriation, social use, significance of use, technique socialization, identity construction, representations, technical imaginary.

Table 5.1. *Comparative analysis of the three approaches of ICT uses*

Our presentation of the different approaches of use corresponds to the way research projects were grouped and also reveals the dominating trends as they appeared recently. The reader understands that the order of presentation of these three trends is sequential (Table 5.1). The diffusion approach is focused on observing and explaining the disparities in terms of the rate of equipment and practice; the innovation approach involves the social dimension of the technical innovation process; and the appropriation approach studies the processes of usage and practice structure. This analysis leads to a major observation: in most cases, ICTs go together with current evolutions to reinforce and accentuate them. Clearly, it is the same for GITs (see Chapters 1 and 2).

5.6. Conclusion

The goal of this chapter was to give a state of the art of works carried out on the diffusion and adoption problem for GITs. The previous reflections enabled us to present models and theories first developed by sociologists, information systems specialists, communication science specialists, etc. in the field of IT in the broadest sense of the term. Although this state of the art is based on studies that were not developed for GITs, in particular, or put in perspective in relation to the few studies carried out in the field of GI, it still shows to what extent the reflections from "sociology of uses" are relevant to understand the specific case of GITs.

It is a fact that GITs, ICTs, uses and espouse rarely corroborate, but there is nothing more efficient for understanding the effects, adoption and use methods than to question real-life practices. GITs are no more than what their users want them to be, regardless of their technical capabilities. They are the result of a social construction process, differing from IT by its deeper spatial anchoring. Their level of appropriation, for example, is greatly influenced by the level of compatibility between cognitive spatial representation and physical spatial representation. The geographical dimension, in its widest acceptance, is the basis of the particular character of GITs with regard to this diffusion and adoption problem.

5.7. Bibliography

[AIT 93a] H. AITKEN, "Science, technique et économie, pour une problématique de la traduction", *Réseaux*, vol. 60, 61-85, 1993 (translation of a part of the work from Aitken published in 1985 by Princeton University Press, USA).

[AKR 93a] M. AKRICH, "Les formes de la médiation technique", *Réseaux,* vol. 60, 87-98, 1993.

[AKR 93b] M. AKRICH, "Les objets techniques et leurs utilisateurs. De la conception à l'action", *Raisons Pratiques*, vol. 4, 35-57, 1993.

[ARG 78] C. ARGYRIS, D. SCHON, *Organizational Learning: a Theory of Action Perspective,* Reading, MA, Addison-Wesley, 1978.

[ARG 95] C. ARGYRIS, D. SCHON, *Organizational Learning II: Theory, Method and Practice*, Reading, MA, Addison-Wesley, 1995.

[BAC 02] D. BACHY, *Développement d'un outil de description et d'analyse des représentations géospatiales dans une organisation*, Memoir from MSc, Department of geomatic sciences, Laval University, Quebec, Canada, 2002.

[BAR 96] T. BARDINI, "Changement et réseaux socio-techniques: De l'inscription à l'affordance", *Réseaux*, vol. 76, 126-155, 1996.

[BEL 92] P. BÉLANGER, *Le système Vidéoway – Les Modalités d'adoption d'un Système Interactif de Télévision*, Doctoral thesis, Department of communication, Montréal University, Canada, 1992.

[BER 82] P. BERNOUX, *Un travail à soi*, Toulouse, Privat, 1982.

[BOU 89] D. BOULLIER, "Du bon usage d'une critique du modèle diffusionniste: discussion-prétexte des concepts de Everett M. Rogers", *Réseaux*, vol. 36, 31-51, 1989.

[BOU 94] D. BOULLIER, "Construire le téléspectateur: Récepteur, consommateur ou citoyen", in A. VITALIS (ed.), *Médias et nouvelles technologies. Pour une socio-politique des usages*, Rennes, Apogée edition, 1994, 63-74.

[BRE 89] P. BRETON, S. PROULX, *L'explosion de la communication*, Paris/Montreal, La Découverte/Boréal, 1989.

[CAL 86] M. CALLON, "The sociology of an actor-network: The case of the electric vehicle", in M. Callon, J. Law, A. Rip (eds.), *Mapping the Dynamics of Science and Technology: Sociology of Science in the Real World*, Basingstoke, Macmillan, 1986, 19-34.

[CAM 95] H. CAMPBELL, I. MASSER, *GIS and Organizations: How Effective are GIS in Practice?*, UK, Taylor and Francis, 1995.

[CAM 96] I. MASSER, H. J. CAMPBELL, M. CRAGLIA (eds.), *GIS diffusion: the adoption and use of geographical information systems in local government in Europe*, CRC Press, 1996, 238 pp.

[CAM 97] H. CAMPBELL, "Institutional consequences of the use of GIS", in P. LONGLEY, M. GOODCHILD, D. MAGUIRE, D. RHIND (eds.), *Geographical Information Systems,* New York, John Wiley, vol. 2, 1997, 621-631.

[CAR 85] A. CARON, H.L. GIROUX, S. DOUZOU, "Diffusion et adoption des nouvelles technologies: Le micro-ordinateur domestique", *Canadien Journal of Communication*, vol. 11, 369-389, 1985.

[CAR 01] C. CARON, S. ROCHE, "Vers une typologie des perceptions spatiales", *L'Espace Géographique*, vol. 1, 1-12, 2001.

[CER 80] M. CERTEAU (ed.), *L'invention du quotidien*, Paris, rUGE, 1980.

[CHA 92] P. CHAMBAT, "Technologies à domicile", *Esprit*, vol. 186, 99-112, 1992.

[CHA 94a] P. CHAMBAT, "NTIC et représentation des usagers", in A. VITALIS (ed.), *Médias et Nouvelles Technologies. Pour une Socio-politique des Usages*, Rennes, Apogée edition, 1994, 45-59.

[CHA 94b] P. CHAMBAT, "Usages des technologies de l'information et de la communication", *Technologies et Société*, vol. 6, 249-270, 1994.

[FLI 95a] P. FLICHY, "L'action dans un cadre sociotechnique. Comment articuler technique et usage dans une même analyse?", in J.-G. LACROIX, G. TREMBLAY (eds.), *Les Autoroutes de l'information, un Produit de la Convergence*, Sainte-Foy, Presses de l'Université du Québec, 1995, 405-433.

[FLI 95b] P. FLICHY, *L'innovation Technique. Récents Développements en Sciences Sociales vers une Nouvelle Théorie de l'innovation*, Paris, La Découverte, 1995.

[GIB 77] J.J. GIBSON, "The theory of affordances", in R. SHAW, J. BRANSFORD (eds.), *Perceiving, Acting, and Knowing. Towards an Ecological Psychology*, Hillsdale, NJ, Lawrence Erlbaum, 1977, 67-82.

[HAR 97] F. HARVEY, "Agreeing to disagree: The social construction of geographic information technology", *Proceedings from the GIS/LIS '97 Conference*, 1997, 808-815.

[HAR 98] F. HARVEY, N. CHRISMAN, "Boundary objects and the social construction of GIS technology", *Environment and Planning A*, vol. 30, 1683-1694, 1998.

[JOU 93a] J. JOUËT, "Pratiques de communication et figures de la médiation", *Réseaux*, vol. 60, 99-120, 1993.

[JOU 93b] J. JOUËT, "Usages et pratiques des nouveaux outils de communication", in L. Sfez (ed.), *Dictionnaire Critique de la Communication*, Paris, PUF, 1993.

[KAT 71] E. KATZ, "The social itinerary of technical change: Two studies of the diffusion of innovations", in W. SCHRAMM, D. ROBERTS (eds.), *The Process of Effects of Mass Communication*, University of Illinois Press, 1971.

[KRO 23] A.L. KROEBER, *Anthropology: Culture Patterns and Processes*, New York, Harcourt, Brace & World, 1923.

[LAC 93] J.-G. LACROIX, G. TREMBLAY, G. PRONOVOST, "La mise en place de l'offre et la formation des usages des NTIC. Le cas de Vidéoway et de Télétel", *Cahiers de Recherche Sociologique,* vol. 21, 80-122, 1993.

[LAC 94] J.G. LACROIX, "Entrez dans l'univers merveilleux de Vidéoway", in J.-G. Lacroix, G. Tremblay (eds.), *De la Télématique aux Autoroutes Électroniques. Le Grand Projet Reconduit*, Sainte-Foy/Grenoble, Presses de l'Université du Québec/Presses Universitaires de Grenoble, 1994, 137-162.

[LAT 89] B. LATOUR, *La Science en Action*, La Découverte, Paris, 1989.

[LEG 03] P. LEGRIS, J. INGHAM, P. COLLERETTE, "Why do people use information technology? A critical review of the technology acceptance model", *Information and Management*, vol. 40, 191-204, 2003.

[LEV 90] P. LEVY, *Les Technologies de L'intelligence*, Paris, La Découverte, 1990.

[MAL 94] P. MALLEIN, Y. TOUSSAINT, "L'intégration sociale des TIC: Une sociologie des usages", *Technologie de L'information et Société*, vol. 6, 315-335, 1994.

[MAS 93] I. MASSER, H.J. ONSRUD (eds.), *Diffusion and Use of Geographic Informations Technologies*, Dordrecht, Kluwer, 1993.

[MAS 95] H. CAMPBELL, I. MASSER, *GIS and Organizations: How Effective are GIS in Practice?*, UK, Taylor and Francis, 1995.

[MAS 96] I. MASSER, H. CAMPBELL, M. CRAGLIA, *GIS Diffusion: The Adoption and Use of Geographical Information Systems in Local Government in Europe*, UK, Taylor and Francis, 1996.

[MIL 98] F. MILLERAND, *Usages des NTIC: Les Approches de la Diffusion, de L'innovation et de l'appropriation*, Doctoral thesis in communication, Montreal University, 1998.

[MOR 86] D. MORLAY, *Family Television, Cultural Power and Domestic Leisure*, London, Routledge, 1986.

[MOR 92] D. MORLAY, *Television, Audience and Cultural Studies*, London, Routledge, 1992.

[NOR 94] D.A. Norman, "Les artefacts cognitifs", *Raisons Pratiques*, vol. 4, 15-34, 1994.

[PER 89] J. PERRIAULT, *La Logique de l'usage. Essai sur les Machines à Communiquer*, Paris, Flammarion, 1989.

[PRO 83] S. PROULX., "Une lecture de l'oeuvre de Michel de Certeau: L'invention du quotidien, paradigme de l'activité des usagers", *Communication*, vol. 15, 171-197, 1983.

[PRO 94] G. PRONOVOST, "Médias: Éléments pour l'étude de la formation des usages", *Technologie de L'information et Société*, vol. 6, 377-400, 1994.

[PRO 95] S. PROULX, M.-F. LABERGE, "Vie quotidienne, culture télévisuelle et construction de l'identité familiale", *Réseaux*, vol. 70, 121-140, 1995.

[ROC 96] S. ROCHE, C. CARON, Y. BÉDARD, "Vers une approche plus complète du rôle de la géomatique dans les organisations", *Revue Internationale de Géomatique*, vol. 6, 73-92, 1996.

[ROC 97] S. ROCHE, Y. BÉDARD, "L'appropriation sociale des technologies de l'information géographique: Quelles leçons pour la mise en œuvre des SIG?", *Revue Internationale de Géomatique*, vol. 7, 297-316, 1997.

[ROC 98] S. ROCHE, "L'appropriation sociale des technologies de l'information géographique", *L'Espace Géographique*, vol. 4, 317-327, 1998.

[ROC 99a] S. ROCHE, J.B. HUMEAU, "La diffusion spatiale des technologies de l'information géographique en France", *Mappemonde*, vol. 53, 20-25, 1999.

[ROC 99b] S. ROCHE, J.B. HUMEAU, "GIS development and planning collaboration: A few examples from France", *URISA Journal*, vol. 11, 5-14, 1999.

[ROC 00] S. ROCHE, *Les enjeux sociaux de Systèmes d'Information Géographique, Les cas de la France et du Québec*, Paris, L'Harmattan, 2000.

[ROD 02] O. RODRIGUEZ, S. ROCHE, F. BROCHU, C. CARON, J.J. CHEVALLIER, "Etablissement de la valeur d'une infrastructure d'information géospatiale selon la perspective de l'interactionnisme sociale", *Proceedings from the Symposium sur la Théorie, les Traitements et les Applications des Données Géospatiales*, Ottawa, Canada, July 9–12, 2002,

[ROG 83] E.M. ROGERS, *Diffusion of Innovations*, New York, Free Press, 1983.

[ROG 93] E.M. ROGERS, "The diffusion of Innovations Models", in I. MASSER, H.J. ONSRUD (eds.), *Diffusion and Use of Geographic Information Technologies*, Dordrecht, Kluwer, 1993, 9-34.

[ROG 89] J.U. ROGGE, "The media in everyday family life: Some biographical and typological aspects", in E. SEITER, H. BORCHERS, G. KREUTZNER, E.-M. WARTH (eds.), *Remote Control, Television, Audiences, and Cultural Power*, London, Routledge, 1989, 168-179.

[SAM 94] C. SAMMER, Y. COMBES, "Évolutions structurelles: des espaces temps revisités à travers l'étude de pratiques d'outils mobiles en milieu professionnel", *Technologie de L'information et Société*, vol. 6, 353-374, 1994.

[SAI 99] R.A.F. SAINT LAURENT (ed.), "Informatique en réseau et nouvelle organisation de la production. Le cas Ouest-France", *Proceedings from the symposium Usages et Services des Télécommunications*, Arcachon, France, June 7–9, 1999.

[SAN 94] L. SANTERRE, "L'informatisation de la production et le bouleversement des modes de vie: des rapports d'interdépendance", *Technologie de L'information et Société*, vol. 6, 337-359, 1994.

[SCA 94] V. SCARDIGLI, "Déterminisme technique et appropriation culturelle: évolution du regard porté sur les technologies de l'information", *Technologie de L'information et Société*, vol. 6, 299-314, 1994.

[SIL 92] R. SILVERSTON, E. HIRCH, D. MORLEY, "Information and communication technologies and the moral economy of the household", in R. SILVERSTONE, E. HIRSCH, D. MORLEY (eds.), *Consuming Technologies. Media and Information in Domestic Spaces*, New York, Routledge, 1992, 15-31.

[SIL 89] R. SILVERSTON, D. MORLEY, A. DALHBERG, "Families, technologies and consumption: The household and information and communication technologies", *ESRC Program on Information and Communication Technologies Conference*, Brunel University, 1989.

[TOU 92] Y. TOUSSAINT, "La parole électrique. Du minitel aux nouvelles 'machines à communiquer'", *Esprit*, vol. 186, 127-139, 1992.

[VED 94] T. VEDEL, "Sociologie des innovations technologiques des usagers: introduction à une socio-politique des usages", in A. VITALIS (ed.), *Médias et Nouvelles Technologies. Pour une Socio-politique des Usages*, Rennes, Apogée éditions, 1994, 13-43.

[VIT 94] A. VITALIS, "La part de citoyenneté dans les usages", in A. VITALIS (ed.), *Médias et Nouvelles Technologies. Pour une Socio-politique des Usages*, Rennes, Apogée editions, 1994, 35-44.

[WOO 91] S. WOOLGAR, "Configuring the user", in J. LAW (ed.), *A Sociology of Monsters: Essays on Power, Technology and Domination*, London, Routledge, 1991, 57-102.

Chapter 6

GIS Engineering: Its Place in Organizations

6.1. Introduction

The main goal of this chapter is to give a state of the art on the problem of GIS development in organizations. The goal is not only to make a technical or methodological description of GIS development methods, but also to give a global overview on questions raised by the implementation in different types of organizations and to show the implications of these implementation methods, as well as the factors influencing them.

Once we have presented the vision of "GIS, project of the organization" that was prevalent for a long period of time, we will show the gap between the current vision of GIS implementation processes and the reality in organizations. Comparing some organizational sociology concepts with this reality will enable us to confirm the consistency of situations encountered and to propose another vision of GIS development in organizations. In conclusion, we will address some questions still being debated.

Chapter written by Henri PORNON.

6.2. GIS, project of the organization

For a long time, GIS implementation processes in organizations were described as the implementation of the organization's project. With the help of computer specialists, GIS specialists or external consultants, the organization expressed objectives through decision makers, then with an appropriate method, would deploy GIS tools to help it reach its objectives in a coordinated and harmonious way (Figure 6.1).

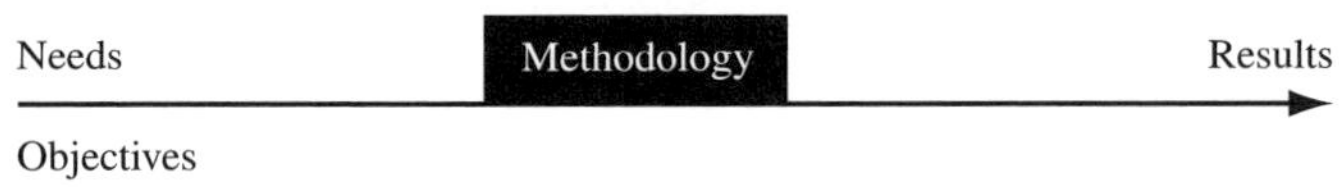

Figure 6.1. *GIS, project of the organization*

The formulation proposed in [THE 96] describes this widespread vision of GIS implementation in organizations: "the harmonious integration of a SIRS in a corporate structure or a public organization involves a planning process that far exceeds the simple integration of data and software. A rigorous design methodology must be followed to reach compatibility between the needs of the institution and the SIRS results".

GIS is the subject of these methods and attempts to federate GI use around a database and a single tool that would guarantee a consistent methodology.

Saindon *et al.* [SAI 92] showed, in the more general context of IT, that this vision is based on a mechanistic perspective for organizations, in which the computer specialist (or the GIS specialist) applies his neutrality in the organization (as a stakeholder) and considers himself the carrier of the organization's objectives and is invested in the responsibility for improving coordination.

6.3. A shift between words and reality

6.3.1. *Observations from GIS specialists*

From 1995, several academic research studies emphasized the shift between this vision of GIS deployment and reality in organizations.

From a survey done in local communities in the United Kingdom, Heather CAMPBELL [CAM 95] highlighted the existence of many power struggles surrounding GIS (especially between computer specialists and users) and the existence of three types of projects corresponding to different approaches:

– The "classically corporate" approach corresponds to traditional understanding of a corporate approach. A large number of services, and maybe even the organization itself, is part of the project under the direction of the IT or urban planning department.

– The "theoretically or pragmatically corporate" approach characterizes an approach that does not correspond to the traditional definition of such an approach even though it involves many departments. It starts with a GIS request from different departments or from an attempt by the computer department to improve coordination in the organization.

– The "fiercely independent" approach characterizes the introduction and development of GIS in a single department. Organizational problems are not addressed.

This study reveals for the fist time the existence of achievements other than organizational dimension projects often presented and contradicts the premise that GIS projects are inevitably or obviously corporate projects with federating objectives. By describing two corporate approaches, she established a distinction between actual organizational approaches (driven by the hierarchy) and approaches that claim to be organizational but are only driven by departments in the organization.

Carrie Anderson [AND 96] raises the question whether the deployment of GISs in organizations can be seen as a project (involving the execution of an action plan) or as the development of a process (continuous development of a technology integrating many changes). Relying on prior studies by PINTO and ONSRUD, she presents two models describing the process of GIS adoption:

– *Content model*: organizational and institutional factors facilitating or restricting the implementation of GIS: people, the organization, goals, change and technology.

– *Process model*: key steps in the implementation focused on technology and the tasks to accomplish.

The criticism here involves the method (project or the process) instead of purposes and objectives.

The book *Geographic Research: Bridging the Atlantic* published in 1997 includes several contributions also questioning the criteria of success for GIS [PIN 97], the role of the "champion" [AZA 97] or organizational parameters [NED 97]. Pinto and Onsrud [PIN 97] notably mentions the fact that there can be winners and losers behind the success of GIS and reminds us of the necessity of considering the political parameters in the organization.

The three doctoral theses [CAR 96], [POR 98a] and [ROC 00] highlight other aspects of the gap between this vision of GIS implementation and organizational reality: power struggles, concurrent projects, "clandestine" SIRS that we forget to talk about, improvized methods, projects justified after the fact, etc.

Caron *et al.* [CAR 96] show that "these technologies do not always contribute to reaching the corporation's formal objectives and that a project never goes the way it is formalized by development methods". Project development is often chaotic and there are often differences between the official and unofficial characteristics of these projects.

In [POR 98a], we show that GIS projects can be departmental (one department in the organization), organizational (involving the corporation) or inter-organizational (common to several organizations). We also learn that two logics of action coexist corresponding to different visions of the usefulness of GIS and to contradictory stakeholder strategies: some projects are aimed at improving coordination in the organization; others are aimed at reinforcing the autonomy of the stakeholders involved.

Roche [ROC 00] highlights the close links between the perception of community stakeholders and the way they use geomatic technologies. These perceptions are related to the training and function of the stakeholders in the municipal organization and to resulting spatial representations. Another criterion involves the level of technological awareness of the stakeholders. He also emphasized the fact that perception by the stakeholders of potential effects from the diffusion of geomatic technologies on their organizational situation also influences the perception of the tools and their level of use and appropriation of these technologies.

Finally [CAR 02], going back to the official motivations for implementing GIS (do more with less, accelerate data collection, facilitate access to data, etc.) also mentions the commonly seen unofficial motivations (spend budget money before a certain date, use a government grant, reduce the number of employees, have more prestige by using high-tech technologies, follow current trends). Mourlon and Neyer [MOU 02] make a similar observation in another domain by analyzing the way Enterprise Resource Planning (ERP) software is introduced in organizations.

Caron and Bedard [CAR 02] show that among the project triggers are opportunity (now is the time to acquire an ICT), problem (gap between what exists and what we want) or even crisis (important problem) and identifies four types of geomatic technology implementations, presented by increasing organizational involvement:

– *isolated GIT*: projects focused on a few specific tasks, GIT is implemented independently by different departments in the organization. These projects only require the modification of a few tasks in the organization;

– *combined GIT*: at least two GITs are combined in a workflow. These projects require that a few tasks be redesigned in the organization;

– *GIS*: managerial considerations are more important than the aspects relating to hardware and software. These projects require rethinking several organizational processes in a managerial perspective;

– *global geography-enabled organizational solution*: the implementation of GITs is integrated to global organizational strategy involving the corporation. GITs are the trigger of a major reorganization or important organizational process support.

These different typologies illustrate the variety of contexts and situations of geomatic tool implementation.

6.3.2. *The analysis by sociologists of computerization processes*

These contributions are sometimes based on studies from several sociologists from organizations questioning the capacity of organizations to express objectives or emphasizing paradoxical situations and the absence of technological or social determinism in organizations [MIN 90], [CRO 77], [MOR 89], [GEN 92].

Other sociologists, specifically involved in the analysis of computerization processes in organizations, show the universality of conflicting situations and the problem with federating the stakeholders around common objectives and projects [ALT 85], [BOU 90], [PAR 89].

6.3.3. *GIS deployment: smooth sailing or chaotic process?*

All these contributions first lead to contest the idea that GIS implementation can be "smooth sailing", where users and computer specialists walk "hand in hand", after the objectives were assigned (by whom?) and an action plan was defined. The diversity of organizational situations involves the objectives assigned to projects and the perception of GIS usefulness, the behavior of the stakeholders and the methods implemented.

Can we still conclude that "GIS development is a chaotic process"? [CAR 02]. We have shown in [POR 98a] that these situations only seem inconsistent if we were to examine them in the mechanistic perspective described in [SAI 92]. Conversely, they become consistent when we analyze them through organizational sociology concepts.

6.4. A few useful concepts of organizational sociology

6.4.1. *Strategies of stakeholders, action and geomatic systems*

In [CRO 77], the stakeholder in the organization is a strategist with a margin of action (zone of control) because he controls a zone of uncertainty. This stakeholder has a rational behavior, in terms of the strategy pursued. Power is a relation of exchange that is negotiated and therefore is not an attribute. In [CRO 77] four resources of power were identified: the rule, expertise, relations with the environment, command of information flows. The interaction between the stakeholders leads to the constitution of a generally stable action system that the strategic analysis strives to update. The models of organization are always contingent.

The study by Crozier and Friedberg [CRO 77] leads us to reconsider the role of the stakeholder in the organization by developing a political metaphor [MOR 89], in which the stakeholder pursues his objectives and generally attempts to reconcile them with those of the other stakeholders in the organization. This metaphor is a very efficient analysis grid to understand the situations of conflicting confrontation generated by geomatic projects. Behind the apparently erratic behaviors of certain stakeholders (competitive projects, "clandestine" projects, opposition to GIS projects, lack of involvement, etc.), in all cases, an analysis of the situation and a strategic behavior are underlying leading them to withdraw or oppose projects that they perceive as going in a direction that is contrary to their own objectives within the organization.

In addition, [CRO 77] shows that the relevant context for the study of phenomena and organizational situations is not always the organization. The concept of action system adopts an intra-organizational (action system limited to a department in the organization), organizational or inter-organizational (action system common to several organizations or to several stakeholders of multiple organizations) point of view.

We have shown in [POR 98a] that the action system is actually a more relevant frame of work to observe the implementation of GIS development and more generally geomatic projects. There are GISs in action systems:

– departmental (the case of management or a department in the company that implements a GIS),

– organizational (the GIS project is aimed at covering all the needs of the organization),

– inter-organizational (stakeholders from multiple organizations get together to implement a GIS or a common geomatic project).

6.4.2. *Computerization process: stakeholders searching for opportunities*

In [ALT 85] and [BOU 90], the processes of introducing technical tools in organizations are analyzed. Stakeholders search for opportunities to improve their situation in the organization (increasing their range of action as given in [CRO 77]). New technical tools (geomatic technologies) offer opportunities and are implemented in real life by these stakeholders in action systems with the support of other stakeholders despite their opposition. In terms of stakeholder strategies, GISs compete with other geographic computer tools (automated mapping, business applications, etc.).

With the study of the introduction of micro-computer technology at RATP, [BOU 90] highlights several stages that we could describe in a similar way as with GIS:

– cognitive learning: isolated individuals experiment a new technology that is developing in the organization;

– restricted groups (within the same office or department): learning new roles and new relations when micro-computers arrive;

– the organization implements what looks from the outside as an orchestrated strategy.

In [BOU 90], the seemingly harmonious character of the progression observed from the outside, in the structure of the micro-computer field, contrasts with the relative "chaos" characterizing it when we look at it from inside is noted.

6.4.3. *Differentiation GIS and integration GIS*

Lawrence and Lorsch [LAW 89] shows that organizations adapt to the constraints of their environment by a double movement of differentiation and integration: the differentiation comes from "differences of cognitive and affective orientation of contexts in the different functional departments", and the integration is the "quality of the collaboration", which exists between the departments that must combine their efforts to satisfy the demands of the environment. We have shown in [POR 98a] how these differentiation and integration forces are active in the geomatic context.

The strength of differentiation is active in "sectoral" GISs, often implemented by operational departments to cover their specific needs, regardless of coordination objectives with other departments in the organization. These departments look at their business viewpoint (urban planning, networks, green spaces, road system, etc. in communities). They strive to have the most adapted business application for their sector, in which they want to manage (retain) their data and the specialization criteria of the tool are a priority over the constraints of cooperation with other departments. From the strategic point of view of [CRO 77], the resources in power are mostly technical expertise and data retention from a field.

The force of integration is active in "organizational" or "federating" GISs, implemented by functional (IT, geomatic, sometimes topography or consulting firm) departments favoring the coordination objectives of the specifics of departmental needs. These departments position themselves from the organization's point of view, and want to have a federating and centralizing tool to improve the consistency and quality of information in the organization, and reinforce their control on the information system. Their power resources are control of technology, methods, control of information flow.

In some organizations, only sectoral GISs exist and that causes problems such as multiplicity of applications, redundancy of basic data, inconsistency, lack of coordination, etc. We often wind up at one point or another asking how to coordinate these sectoral applications.

In other organizations, there is only one federating project, which does not always convince users in departments to use it. Organizations that have implemented federating tools and common databases must therefore convince users to use them and sometimes make concessions on the homogenity of the tools or the uniqueness of the data.

Finally, in several organizations, both families of tools coexist (sectoral and federating tools), with or without coordination. Several coordination methods of these two approaches can quickly be brought up:

– IT architectures and concepts involving the coordination of tools in particular: client/server, spatial servers, software interoperability tools, data exchange between applications in standard formats, etc.;

– resource sharing (data storage on servers, use of common peripherals), data sharing (common reference, shared databases), methods of organization (project lead, coordination, job specialization, definition of department prerogatives, etc.);

– incremental or participative methods, negotiated approaches;

– integration of conflict resolution procedures to implementation methods, geomatic strategic management in organizations.

6.5. The development of geomatics in organizations

It is now possible to better understand the development of geomatics in organizations.

6.5.1. *Project or process?*

Following the example [AND 96], we can first mention the development of geomatics as a continuous process, which can take the form of a project implemented with an action plan at certain stages. The loop diagram from [GRA 96] illustrates the cyclical aspect of this development (Figure 6.2).

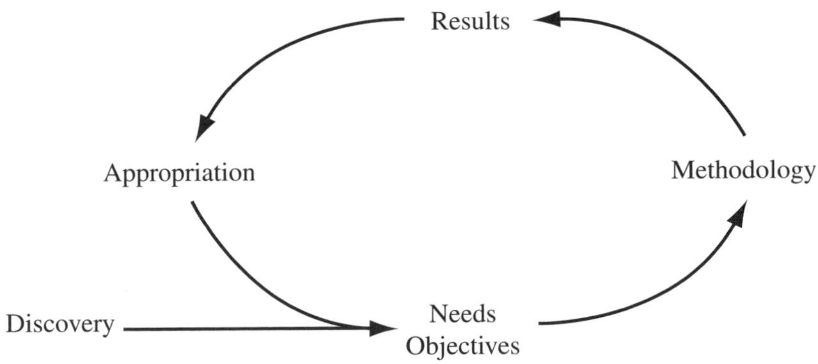

Figure 6.2. *Cyclical GIS, process (source: [AND 96])*

6.5.2. *Projects from stakeholders or from the organization?*

The best way to understand the paradoxical situations in which several stakeholders implement competitive tools or tools that are not inter-connected is to maintain that they are projects from stakeholders in the organization and not projects from the organization.

In this way, when the GIS specialist implements a GIS project with the objective of improving coordination between departments, he plays his role, but still has not received the mandate of federating geomatics around a single tool, even though we can actually hope that this coordination effort will produce a positive effect on the organization. In certain cases, decision makers at this stage are not able to assign objectives because they do not understand technology and its organizational impact.

Similarly, when the manager of an operational department decides to acquire a GIS business application bringing his department more performance and to generate gains in productivity, he has only received an implicit mandate to search for this type of improvement and we can also hope that this performance effort will produce a positive effect on the organization.

6.5.3. *A model to identify the descriptive parameters of the methodologies*

On the basis of [MOR 77] ("Action creates organization creates action"), we can represent the development of geomatics in organizations in the form of a series of action/organization cycles, in which stakeholders (in search of opportunity or reacting to initiatives from other stakeholders) assign their own objectives to actions producing results modifying the organizational context and the development state of geomatics (Figure 6.3).

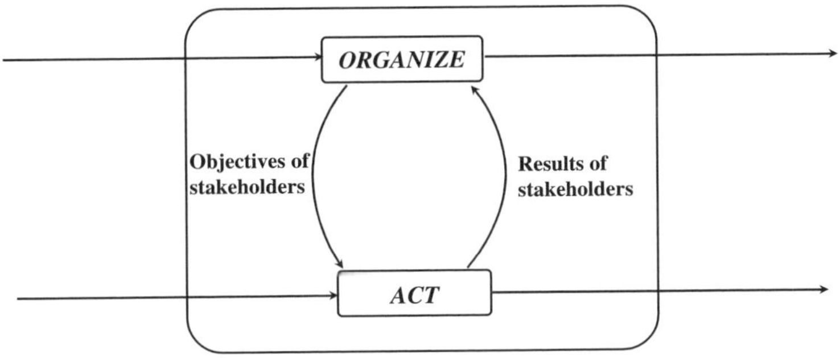

Figure 6.3. *Action/organization cycles (source [POR 98])*

Among the objectives and goals of the stakeholders, we find the official and unofficial motivations described in [CAR 02]: do more with less, accelerate data collection, facilitate access to data, spend budget money before a specific date, use government grant, reduce the number of employees, acquire more prestige by using high-tech technologies, follow current trends, etc.

Among the actions, there are processes of project development steps, as well as reactions: protesting or sabotaging a project (as we have shown in [POR 98a] for the GEOM case), competitive initiative, passive behavior, etc. On the one hand, results are technical (a project, a tool, a consultant report, a database, etc.), and on the other hand, results are strategic: increase or decrease in the importance of supporters and detractors of the organization's action. These results lead to a new organizational context contributing to the trigger of new actions continuing or contradicting the previous initiatives.

This model can be explained in more detail by describing the flows of stakeholders and organizations involved in actions (stakeholders or organizations "enter" and "exit" the field of observation based on actions and projects) and constraints, opportunities and innovations are taken into consideration in the observation process (Figure 6.4).

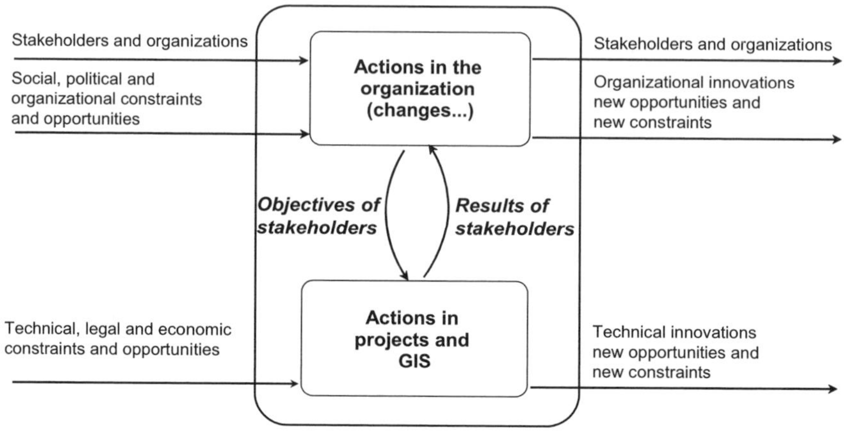

Figure 6.4. *The action/organization model (source: [POR 98a])*

The interest with this model is the identification of action methods for the different stages of geomatic development.

6.5.4. *Synthesis: the process of geomatic development*

To describe the reality of geomatic development processes in organizations, we proposed in [POR 98a] an alternative formulation to the one [THE 96] addressed in section 5.2:

> *The integration of geographic computer tools in an action system is generally done by a series of generally coordinated or competitive initiatives. Each initiative corresponds to the perception that a stakeholder (individual or collective) has the usefulness and interest of technology in the context of his strategy and his role in the action system. The needs and responses to the needs are gradually built during appropriation of the technology by stakeholders. At a certain point, the projects seem to become projects of the organization, because of the emergence of a global strategy and the assignment of objectives to geomatics in managerial terms.*

6.5.5. *Descriptive parameters of organizational contexts*

Can we identify the explanatory parameters of the diversity of these methods and approaches? Without going over an exhaustive list and identifying cause and effect relationships, we can mention:

– the relevant action system: in certain organizations, the absence of relations between executives for example explains the fact that we can find federating intra-organizational GISs (within executive committees); marginal secant stakeholders [CRO 77] from several organizations can, by extending their coordination activities, associate to implement an inter-organizational geomatic project, as federating or sectoral projects are developed in their respective organizations;

– the level of corporate centralization/decentralization: we find sectoral methods in decentralized organizations more easily, and conversely, there are more federating methods in centralized organizations;

– the size of the organization: experience shows that the larger the organization, the more difficult it is to coordinate the different initiatives linked to geomatics;

– individuals and individual behaviors: regardless of organizational constraints, individuals retain margins of action enabling them to influence the course of events;

– the managerial context, notably the priority method of geomatic and IT projects.

– financial and human resources constraints: when the financial situation and constraints of personnel do not enable the consideration of federating projects that are more expensive in terms of investment and presenting more risk of failure, priority is often given to sectoral projects.

6.6. The usefulness of geomatics in organizations and on the field

The review of geomatic development processes in terms of concepts of corporate sociology leads us to question the strategic objectives that we can attribute to geomatics and its usefulness in organizations. We can in fact ask some questions:

– Is it correct to favor federating GIS over sectoral GIS?

– Should we deduce from this analysis situations where improving coordination and developing information sharing are objectives impossible to reach?

– What could be the usefulness of geomatics in organizations?

– What can be the usefulness of GISs on the field?

– How can we measure the organizational impact of GIS?

We quickly review these questions, which are far from being exhaustive on the theme of the organizational usefulness of GISs, enabling us to propose a few paths of reflection.

6.6.1. *Favoring federating or sectoral GISs?*

A quick overview of the history of geomatics leads us to identify three "ages", as with IT. The first age is centralization. In IT, computers are large systems (mainframes) difficult to use, where computer specialists completely dominate the situations, uses and conditions of deployment of centralized IT. Similarly, in geomatics, the high cost and problems with using the first GIS software packages have a centralizing effect; only a few specialists can use these not very "user friendly" tools most often implemented in federating projects.

The second IT age coincides with the arrival of the first micro-computers, with a decentralizing impact enabling users to appropriate IT tools and to free themselves from the control of computer specialists ([ALT 85], [BOU 90]). In geomatics, it is the ARC/VIEW,

MAPINFO, GEOCONCEPT generation, often acquired independently by departments within sectoral projects.

The third IT age (which is not the age of senility, but of maturity) starts with the emergence of client/server concepts that will allow, following the example of other recent concepts, a possible overview of these contradictory logics. Micro-computers communicate with mainframes. In fact, it is no longer possible to identify logic or a type of predominant stakeholder, IT tools that can be deployed in contexts and with very different objectives. Similarly in geomatics, recent evolutions involving spatial servers, Internet environment diffusion Internet tools, GIS over personal digital assistant (PDA), or software interoperability lead to a situation in which tools no longer mandate an organization (centralized/decentralized) or a logic (federating/ sectoral), but can on the contrary adapt to all types of objectives and all forms of organization.

Concerning the question of knowing which organizations have the best performance (the most centralized or the most decentralized), [LAW 89] observed that the states of differentiation and integration are in opposite relation (the more differentiated a company, the harder it is to reach good integration). On the other hand, organizations with the best performance reach the highest degree of integration, and are also the most differentiated.

In this perspective, we can declare that to contribute to making organizations reach better performance, GISs should not be implemented in only federating or sectoral logic, but must use these two logics. In fact, the methods must integrate conflict resolution devices between stakeholders carrying these two logics.

6.6.2. *Cooperation and information sharing: an optimistic objective?*

Are improving coordination and developing information sharing impossible objectives to reach? This question is not insignificant, because a large number of federating GISs justify the important capital committed to their deployment in the promise of achieving such objectives.

It is obvious that certain naive and optimistic approaches, seeming to consider that information is easily shared simply by the deployment of technical tools, are most often contradicted by the acknowledgment that deploying these tools did not lead to improvement on this point. Apparently, it is not a technical problem. However, it would be wrong to adopt the disillusioned and pessimistic point of view of those considering that information retention is the only right behavior in organizations and that these objectives are unrealistic. There are in fact situations of information exchange and sharing in and between organizations.

We therefore suggest adopting a realistic point of view and observing that there are cooperations and situations of formal or informal exchanges in or between organizations, but although facilitated by computer tools, these cooperations and exchanges must be analyzed in a social and relational context. A new computer tool creates a new opportunity for initiative that stakeholders will either appropriate or not based on the context. To share information better and to improve cooperation, we can deploy computer tools, but this deployment must be accompanied by a project methodology with the objective of improving communication and dialog between the stakeholders involved.

6.6.3. *Changing points of view on the use of geomatics in organizations*

For a long time and despite abundant references to systemic concepts, we imagined only the organizational usefulness of GIS from a "mechanistic" point of view (in relation to the metaphor of the organization as a [MOR 89] machine) described in [SAI 92]: centralize, unify, eliminate redundancies, unify flows to reach an ideal state. It seems, however, that sociologists and change specialists in organizations rely on a systemic corporate approach and arrive at completely different conclusions [GEN 92]: IT must lead to a diversification of flows, to a redundancy (albeit controlled), and be designed in a permanent evolution logic. In addition, the conflict must not be considered as dysfunctional, but as a factor of integration. In fact, beyond the aspects already addressed on the centralizing or decentralizing impact of GIS, their organizational usefulness cannot be the unification of flows and data centralization. This "systemic" vision of organizations (Table 6.1) leads us to observe that the idea of information stored only once in a single database is poor judgment, many stakeholders in the organization may need to develop distinct visions of the same objects. In fact, beyond eliminating redundancies involving the "background" and reference data with very little value, redundancy of data corresponds to the diversity of territory perceptions by stakeholders and businesses.

"Mechanistic" point of view	**"Systemic" point of view**
Centralization, unification	Diversification and integration
Elimination of redundancies	Managing redundancies
Unification of flows	Diversification of flows
Priority to coding information	Priority to information significance
Conflict = dysfunction	Conflict for better integration
Avoid the unexpected, ignore the unstable	Accept the unexpected and the unstable
Rigidity	Flexibility
Change = access to an ideal state	Change = permanent evolution

Table 6.1. *Organizational usefulness of GISs: mechanistic and systemic point of view*

6.6.4. *Changing points of view on the usefulness of GISs in the territory*

Are GISs tools intended to unify the vision of the land or to share and compare visions of the land? The metaphor of the millefeuille illustrates this GIS approach with the layering of themes (classes, layers, objects, etc.), it will give the land planner an integrated vision of this territory and make better decisions for the land.

We can first criticize this metaphor by saying that the millefeuille is all the more indigestible and difficult to ingest as the number of layers is great. In addition, is it not illusory to believe that the land planner can interpret by himself the information from the agronomist, forester, economist and the urban developer?

It actually seems more realistic to consider that the GIS is a tool that will allow different stakeholders to compare their point of view on the land, to present them to each other, sometimes may bring them together, as long as their cognitive models accept GIS as a modeling tool for the territory in their field, which is not guaranteed, as Wladimir MAJOR and François GOLAY show in this book (Chapter 7).

6.6.5. *Measuring the organizational impact of GIS*

Can we measure the organizational impact of GISs? It is the objective of the "Monitor" described in [POR 98a] (Figure 6.5). It tracks over time the different "action/organization" cycles in a given action system, it represents the different initiatives from stakeholders and groups of stakeholders from the point of view of their impact in terms of differentiation (autonomy) of integration (coordination), it illustrates the diversity in the situations encountered, it formalizes diagnoses and finally it offers propositions to establish a better balance of GIS achievements in the organization, as we have done in [POR 98b], in the analysis of organizational situations linked to GIS with the French facilities ministry.

An application to manage the initiatives of an action system and to represent them graphically in ACCESS can be downloaded at http://www.ieti.fr/1p_moniteur.htm. We will not present the concepts of the Monitor; we will limit ourselves to presenting an example, the case of a large French community described in [POR 98a].

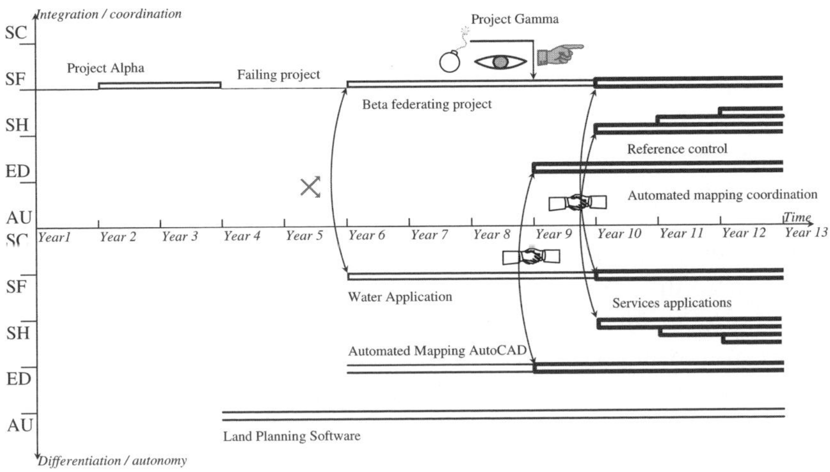

Figure 6.5. *An example of Monitor*

6.7. Conclusion: some questions under discussion

6.7.1. *What roles do geographic computer tools play?*

Until now, we have only discussed geomatic tools (computer tools, databases, etc.) as opportunities and support for strategic initiatives taken by stakeholders in their organizations. This passive vision of their organizational impact overshadows their capacity to produce change, regardless of the organizational context. The first RATP micro-computer users could establish only a balance of power with computer specialists who had complex, expensive computer tools that are difficult to handle because these new computers were easy to use, inexpensive and simple to implement.

Similarly, in the field of geomatics, the IT power pendulum swings to the tune of technological evolutions: gas plants early in the geomatics development forcing organizations to launch centralizing methodologies, office automation GIS followed to help users in implementing their own applications in a decentralizing logic, client/server tools and spatial servers giving the advantage back to computer specialists, etc. Sometimes, the tools have a clear centralizing or decentralizing impact, but in certain cases, it is the way in which users appropriate them that will give them a federating impact or a feeling of autonomy.

6.7.2. *What roles do methods play?*

Until now, we have addressed only project management methods to indicate that some stakeholders may use them as a strategic resource to solidify their role in the organization. This "strategic" vision overshadows the use of methods to formulate the needs, to contain the methods and minimize the risks of failure of the project. We should not judge too quickly that the methods are only there so that computer specialists and GIS specialists can solidify their domination over users. This is all the more true as the methods have evolved over time to become closer to the users. Prototype designs for example have helped in actively involving users with projects and minimized the tunnel effects observed in traditional approaches. However, it seems risky to infer that object methods based on UML formalism, where the impact on the reuse of computer components is indisputable, could have a significant organizational impact on project control. At the most, we can hope that social and technical approaches described in [REE 99] or [EAS 88] create these spaces of negotiation necessary to the resolution of conflicts triggered by geomatic projects.

We can observe that the impact of methods on project management is linked to implementation methods (methods with more or less involvement, formalisms that are not always understandable by users, etc.), as well as the strategic context in which they are used and the stakeholders implementing or experiencing them.

6.7.3. *The paradoxical role of project managers (the champions)*

The North-American myth of the champion [AZA 97] can be examined in terms of this vision of GIS development. Regardless of his devotion to the organization he works for, and his efficiency to the GIS cause in this organization, the champion remains a stakeholder reconciling his personal strategy with his perception of what is right for the organization.

Saindon *et al.* [SAI 92] maintains that computer specialists (and thus GIS specialists) are not neutral, and that they must negotiate their involvement and justify their intervention. We (GIS specialists, consultants, project managers, GIS researchers and other stakeholders in the field) should question ourselves. What kind of IT (geomatics) do we want? What type of stakeholder are we? What is our strategy? What resources of power do we have? Why do we want to implement geomatics in organizations?

We should avoid hiding behind IT methods to avoid social relations, and be conscious of the fact that in terms of geomatics, there is not "*one best way*". Confrontations, contradictions and conflicts are necessary to the introduction of geomatics in organizations. This tricky and sometimes unpleasant introspection leads to other questions on the usefulness of geomatics in organizations that we will not address here.

6.7.4. *Conclusion: transposing to land planning?*

We can offer an analogy between geomatics development processes in organizations and land planning processes: the necessity of admitting that many stakeholders can have contradictory objectives and that struggles of power are inevitable in the territory as in organizations, because of a lack of global rationality that everyone could embrace (the harmonious development of the territory is the product of the same unrealistic vision as the GIS project in the organization to which everyone could adhere to). On the other hand, each stakeholder is rational in his understanding of the challenges and constraints he encounters.

Action systems and stakeholder logic may be easier to identify in the field of land management than in the field of GIS, because political processes are more generally accepted in these situations. Despite this confrontation of strategic stakeholders, the problem comes from the necessity of connecting very sectoral visions (and thus not very integrated) of the territory together: the visions of the economist, environmental or transportation specialists, agricultural specialist, urban planner, etc. This integration of land perceptions in a common perspective raises the same methodological (how can we make these stakeholders communicate together) or even epistemological (is the GIS millefeuille an integrated vision of the territory) problems.

In fact, land planning processes also lead us to give greater importance to involvement and territorial negotiation methods.

6.8. Bibliography

[ALT 85] N. ALTER, *La Bureautique dans l'Entreprise: Les Acteurs de L'innovation*, Paris, Editions Ouvrières, 1985.

[AND 96] C. ANDERSON, "GIS development process: A framework for considering the initiation, acquisition, and incorporation of GIS technology", *URISA Journal*, 8(1): 10-26, 1996.

[AZA 97] B. AZAD, "Leading and managing the GIS implementation process: Beyond the myth of the GIS Champion", in M. CRAGLIA, H. COUCLELIS (eds.), *Geographic Information Resarch: Bridging the Atlantic*, London, Taylor & Francis, pp. 146-164, 1997.

[BOU 90] H. BOUCHIKHI, *Structuration des Organisations: Concepts Constructivistes et Étude de cas*, Paris, Economica, 1990.

[CAM 95] H. CAMPBELL, I. MASSER, *GIS and Organizations: How GIS Are Effective in Practice*, London, Taylor & Francis, 1995.

[CAR 96] C. CARON, *Cadre Descriptif des Projets D'implantation de Technologies Géomatiques dans les Organisations*, Doctoral thesis, Centre de Recherche en Géomatique, Laval University, Quebec, 1996.

[CAR 02] C. CARON, Y. BEDARD, "Lessons learned from case studies on the implementation of geospatial information technologies", *URISA Journal*, vol. 14, No.1 2002.

[CHE 02] J.J. CHEVALLIER, C. CARON, "Développement de projets en géomatique: Du déterminisme technologique à l'approche participative", *Revue Internationale de Géomatique*, Hermès, Paris, vol. 12, no. 4, pp. 489-51, 2002.

[CRO 77] M. CROZIER, E. FRIEDBERG, *L'acteur et le Système*, Paris, Le Seuil, 1977.

[EAS 88] K. EASON, *Information Technology and Organizational Change*, London, Taylor & Francis, 1988.

[GEN 92] D. GENELOT, *Manager Dans la Complexité*, Paris, INSEP Editions, 1992.

[GRA 96] A.D. GRAAFLAND, "Development and adaptation of electronic networks for geo-information processing in municipalities," *Proceedings from JEC*, Barcelona, 1996.

[LAW 89] P. LAWRENCE, J. LORSCH, *Adapter les Structures de L'entreprise*, Paris, Editions d'Organisation, 1989.

[MAS 96] I. MASSER, F. SALGE, *GIS Diffusion: The Adoption and Use of Geographical Information Systems in Local Government in Europe*, London, Taylor & Francis, 1996.

[MIN 90] H. MINTZBERG, *Voyage au Centre des Organisations*, Paris, Editions d'Organisation, 1990.

[MOR 77] E. MORIN, *La Méthode, Tome 1: La Nature de la Nature*, Paris, Le Seuil, 1977.

[MOR 89] G. MORGAN, *Images de L'organisation*, Quebec, Presses de l'Université Laval, 1989.

[MOU 02] S. MOURLON, L. NEYER, "Tout ce que nous avons voulu savoir sur les ERP", memoir from student engineers from Ecole des Mines de Paris, 2002.

[NED 97] Z. NEDOVIC-BUDIC, "GIS technology and the organizational context: Interaction and adaptation", in M. CRAGLIA, H. COUCLELIS (eds.), *Geographic Information Resarch: Bridging the Atlantic*, London, Taylor & Francis, pp. 165-184, 1997.

[PAR 89] F. PARVÉ, *L'illusion Informaticienne*, Paris, L'Harmattan, 1989.

[PIN 97] J. PINTO, H. ONSRUD, "In search of the dependant variable: Toward synthesis in GIS implementation research", in M. CRAGLIA, H. COUCLELIS (eds.), *Geographic Information Resarch: Bridging the Atlantic*, London, Taylor & Francis, pp. 129-145, 1997.

[POR 97] H. PORNON, Géomatique et Organisations: Contradictions et Intégration des Projets D'acteurs, Doctoral thesis, Ecole Polytechnique Fédérale de Lausanne, 1997.

[POR 98a] H. PORNON, *SIG, Pouvoir et Organisations: Géomatique et Stratégies D'acteurs*, Paris, L'Harmattan, 1998.

[POR 98b] H. PORNON, *SIG, et Organisations*, study report published by CERTU (Ministère de l'Equipement), Lyon, 1998.

[POR 98c] H. PORNON, *Partenariats Relatifs aux SIG*, study report published by CERTU (Ministère de l'Equipement), Lyon, 1998.

[ROC 00] S. ROCHE, *Les Enjeux Sociaux des Systèmes D'information Géographique*, Paris, L'Harmattan, 2000.

[REE 99] D.E. REEVE, J.R. PETCH, *GIS, Organisations, and People: A Socio-technical Approach*, London, Taylor & Francis, 1999.

[SAI 92] M. SAINDON, M. LANDRY, C. BANVILLE, *Pour un Élargissement de la Conception du Changement Informatique*, off-printed, Laval University, Quebec, 1992.

[THE 96] M. THERIAULT, *Systèmes d'Information Géographiques: Concepts Fondamentaux*, Laval University, Quebec, 1996.

Chapter 7

GIS, Cognition and Business

7.1. Introduction

To create a city, or more generally, an urban or rural space, institutional stakeholders coordinate their actions to complete land projects, for example, creating roads or communication lines to help in the accessibility of a district or a region, or for the development of economic activity and housing, to rebuild a district with coordination, or to give a "cultural light" to memorable areas by preserving the elements defining the history of a population and its territory.

Institutional stakeholders have business practices and, in a certain measure, ideologies; their focus is first a representation of their territory, and second, based on elements of the territorial project, coordinating with their partners within a process of dialog. Some methods are used to obtain active involvement from other stakeholders in this process, for example, citizens involved in the project [CRA 98].

What is the role of GISs in the practice of territory trades? Can we assemble all the knowledge of these trades into a single GIS? Or more

---

Chapter written by Wladimir MAJOR and François GOLAY.

importantly, can we use these trades as a federating element? What are the cognitive models of territorial stakeholders made up of? Are they compatible in their different dimensions? And for which types of collaborations? These are questions that we must ask to understand the possible use of GISs in the practice of territorial businesses.

This chapter addresses these questions and gives some responses. To do this, in the first section, this chapter relies on systemic and sociologic knowledge of organizations to clarify the nature and challenges of territorial cooperation. In the second part (section 7.3), territoriality is presented as emerging from cognitive worlds inherent to each stakeholder, or in more precise terms, to each group of stakeholders within the same *profession*. The diversity of cognitive business worlds is presented next (section 7.4), from a lexical analysis of characteristic documents created by the main businesses of land development (e.g. the Swiss canton of Geneva). Lastly (section 7.5), the limits of GIS are emphasized by way of this diversity, and a few solutions to these limits are proposed.

7.2. Nature and challenges of territorial cooperation

7.2.1. *The territory, a challenge for stakeholders striving for power*

The territory is a real challenge because it is first and foremost a resource space. It becomes the support of an action system organized according to the stakeholders involved and the territorial project engaged in.

Clastres [CLA 74, p. 34] maintains that the power is "at [...] vital structural levels in the company, i.e. at the heart of communication." On the other hand, Crozier and Friedberg [CRO 77, p.72] define "power [as] a relation and not an attribute of stakeholders". It is "the scope of the zone of uncertainty that the unpredictability of his own behavior lets him control in relation to his partners".

These two quotes confirm that, in the context of territorial cooperation, power struggles occur in the negotiation of the territorial project, thus being at the heart of the cooperation process dynamic.

Pornon [POR 97, p. 30] clarifies the nature of the accompanying power play: "These conflicts are triggered by the lack of resources on the territory. They are also caused by the different ways the territory will be used or how social systems will work, and the necessity to coordinate the stakeholders in organizations and on the territory."

This confirms what Ausloos maintains [AUS 95, p. 55] in another context, that "in fixed transaction systems," which is often the case with territorial projects contained by a normative regulation of interventions, "power is in the game as long as it is not really in the hands of the one representing it, but in the tight network of often contradictory networks and resulting compromises."

We understand the importance of clearly understanding what territorial cooperation represents in terms of power play and will highlight possible challenges.

The territorial project's main object is to bring up a problem that the territorial cooperation must solve. To do this, it must have the best methods of representation, and that is where GISs come into play. Each stakeholder will express his or her vision of the territory and objects that seem representative of the project or the problem presented. Real challenges are revealed in the context of cooperation.

7.2.2. *Rules of territorial cooperation*

The goal of territorial cooperation is to settle possible conflicts in land management. It operates by searching for a consensus on actions to carry out their coordination. Territorial cooperation, for example, is one of the regular management techniques for establishing an overall development plan. This cooperation usually relies on principles set up by a specific law.

Territorial cooperation, as a means of concrete expression by the stakeholders, is based on three features:

– *Recognition of the status and role of the people involved*, for example, recognizing the official mission of management departments or associations involved in the problem to be solved (unions, neighborhood associations, consumer associations, political parties, chambers of commerce, association of real estate developers, international organization representatives, etc.). This recognition confirms the influence of these stakeholders over territorial decisions and their credibility. This recognition is obviously a direct function of the degree of legitimacy of the stakeholder in the action involved.

– *Definition of a common project involving explicit objectives*, often defined by pre-established principles as part of longer-term strategies, objectives that are similar to the main points of the land management plan or environmental protection laws already enacted.

– *Possibility of referring to common concepts or to information-processing tools for representation and dialog*. These concepts are sometimes written out in specific legislations, as, for example, zones of allocation or distribution between cultivable and fallow areas, or in territorial databases (cadastre and land registers, inventories of natural and artificial obstacles, etc.).

However, because the rules of the game (legislation) are never completely explained and cannot include all possible territorial situations, there is much left to interpretation for the stakeholders involved, as Crozier and Friedberg [CRO 77] and Pornon [POR 97] have shown. This room to maneuver, this game of influence and stakeholder position, opens the door to contingent solutions based on a temporary compromises negotiated by the stakeholders in the context of cooperation. The line of argument used is very important.

7.2.3. *Real challenges of territorial cooperation*

Because territorial cooperation represents a challenge for the stakeholders, a challenge that will influence their actions in the collective project, cooperation can resemble a game.

In the business world, Pornon [POR 97, p. 20] has described the two types of strategies that were described by Lawrence and Lorsch [LAW 89]: "'separatist' objectives and differentiation strategies of the hierarchy and operations center on one hand, coordinating or centralizing objectives and integrating strategies for the technostructure and logistical support."

That is also what Ricoeur [RIC 90, p. 167] maintains when he says "the real nature of the narrative identity is only revealed [...] in the dialectic of ipseity and sameness."

During territorial cooperation, a system of interactions and influences is created, what Schwarz [SCH 94] calls, in the systemic context, "*differentiation – integration forces*", leading to the results of this cooperation.

For the stakeholder, one of the challenges of territorial cooperation is to confirm his or her legitimacy and identity by involvement in the process of territorial project definition. This identity and legitimacy are developed by negotiating with partners in a continual confrontation of interests and representations: identity, otherness, and a community of interests are thus linked in the process of cooperation. In addition, in order to affirm his position, the stakeholder must distinguish himself through discourse, the key principle of which is to argue. King, who strives to establish the role of models in environmental management, goes so far as to maintain [KIN 93] that "if they [models] do not reveal themselves useful as weapons in the political debate, they will play a very little, if any, role in this debate".

Thus, territorial cooperation requires the creation of a space of communication as a meeting place for the stakeholders involved in a common project:

> *"The act of communicating is not expressed by a transfer of information from sender to recipient, but is mutual modeling of a common world with combined actions: our social realization, through language, gives life to our world. There are linguistic actions that we continuously do: affirmations, promises, requests and declarations. In fact, such a continuous network of*

conversations, including their satisfaction conditions, does not constitute a tool of communication, but the real pattern on which our identity is drawn."[VAR 96]

Challenges take shape in the space of communication, where the legitimacy and credibility of stakeholders as well as their identity are defined. Not only that, we believe [MAJ 99] that this identity is defined by a symbolic story always confronted with the views of other stakeholders and one that is always transformed. In a way, we can see that the stakeholder tells himself a story, argues with himself, giving meaning to his vision of the world. He builds some sort of "intrigue" with his identity in the context of the common vision necessary to the project and in accordance with his role. He plays with his identity by developing a specific argument, relying on possible actions and explicit representations to ensure his credibility.

It is important to note that the "common" elements of the project are unconditionally accepted by the group in the context of compromise. This does not mean that there is a single representation of the object involved. On the contrary, it is possible that the group adopts common elements, without explanation, to retain the ambiguity that is necessary to the compromise [LAN 95].

The group must manage what we call "informational slack," in analogy to the notion of "organizational slack" as proposed by Cyert and March in 1963 [CYE 63]. The notion of "organizational slack" is based on the following idea [FRI 93, p. 47]: "the operation of a complex organization is always based on the existence of reserve resources ensuring a minimum of autonomy in relation to its environment, and by which its operation can settle for quasi-solutions, redundancy, sub-optimal resource allocation". Similarly, the operation of negotiation relies on a certain number of semantic resources and uncertainties necessary to the argument of each stakeholder to develop and lead to solutions for a common project.

"Informational slack" is made up of the "unsaid," the ambiguities and uncertainties inherent to the operation of cooperation. It is present in each of the semantic environments of the stakeholders involved.

The more we try to decrease ambiguity and uncertainty, the more we deplete the "informational slack", and the lesser the chance of coming to a common solution because the common elements of the project can no longer be used by each stakeholder in his or her cognitive environment.

Finally, we must note that, in territorial cooperation, rules and standards to support the exchange between partners are created. As Reynaud [REY 97, p.25] reminds us, "the exchange is possible because of the mobilization of production of the rule."

To establish this type of stable state represented by cooperation, partners must also agree on the rules of cooperation and action, and these rules cannot contradict the partners' value system. There is also the creation of a prescriptive environment developed in the context of cooperation that can be materialized, for example, by the implementation of planning tools or through the definition of general principles of action. The methods of GIS representation are a part of this environment.

The credibility and legitimacy of the stakeholders, the negotiation of uncertainties, the management of ambiguity, and the creation of "informational slack" enable each stakeholder to argue for his identity while remaining in tune with the other stakeholders. in These elements in the search for a common solution and for rules and standards developed with the objective of efficiency are all elements that participate in the context of territorial cooperation and will only be developed when each have been address and all stakeholders have come to an acceptable compromise regarding each.

7.3. Territoriality: an emergence of the cognitive environments of the stakeholders

Chapters 9 and 10 show how territorial cooperation is carried out in a context of negotiation and power plays in which stakeholders negotiate their identity and express their vision of the world. This cooperation must allow for the resolution of conflicts that occur and that allow for the management of semantic ambiguity ("informational slack" management). This is, in a way, the sociological and systemic aspects of territorial cooperation.

However, cooperation must involve a real project, organized around concrete territorial objectives. These objectives can be different for each stakeholder concerned. In addition, the stakeholders become involved in the territory through these objectives. They build a series of relations with the territory. These relations constitute what we call territoriality. Because territoriality creates the cognitive environment of the stakeholders, in the sense that it concretely involves territorial objects, we will focus on the relations that the stakeholders maintain with the territorial objective and on the process of the emergence of territoriality.

7.3.1. *The territorial object and territoriality*

Raffestin [RAF 97] confirms our definition of territoriality: "It is through multiple systems of relations to things, territorialities in a way, that men can satisfy their needs. What we commonly call nature, the social representation of the inorganic and organic, and culture, the symbolic series and material produced, constitute the bases and conditions of multiple interactions emerging in natural and human ecosystems."

These relations refer to the diversity of possible territorial resources, real and concrete for physical objects (land, forests, rivers, mountains, etc.), abstract for objects perceived by humans (pastures, buildings, companies, etc.), and symbolic for objects arising out of intentions or norms (boundaries, named places, monuments, etc.).

Territoriality, that is the experience of the stakeholder with the territory through the objects that he or she sees, is built on three dimensions:

– the physical dimension, designating our relationship with the concrete environment;

– the cognitive dimension, designating the processes of object representation and identification;

– the symbolic dimension, considering symbols or signs associated with the territory.

7.3.2. *The territorial object and its social representation*

The object itself has no meaning; its takes its meaning, its quality, through its projection in interactivity, in its comparison with others and through an inter-subjective dimension. In this projection in interactivity, the characteristics of the object as it is recognized by the group are determined.

That is what George Herbert Mead talks about in the field of anthropology (Liliane Pierrot [POC 96, p. 41]): "To identify and understand life in a group, we have to identify its world of objects. This identification must be made in terms of the significance of these objects for the members of this group".

In territorial practice, the object becomes the support of social interaction; because the action involves it, physical networks are created in the sense of Callon and Latour [CAL 91] in which the stakeholders and the physical supports form the links in a chain. But these networks are also the support of social connections that we will qualify as "instrumental proxemic" links because they constitute a space built by the stakeholders, consciously or not, who use objects as instruments in their relation to others.

This is confirmed by Dosse [DOS 95, p. 138] using the idea of Pierre Levy:

> *The object itself cannot be separated from its conditions of emergence and social extensions; it determines nothing by itself, but contributes to creating new possibilities in the communication order. It includes two axes: one, temporal, it is an axis of interpretation, of giving meaning, and the other,*

spatial, is the axis of the transfer of information from one point to the other.

Figure 7.1 provides a schematic representation of the territorial object's role as a support to social relation.

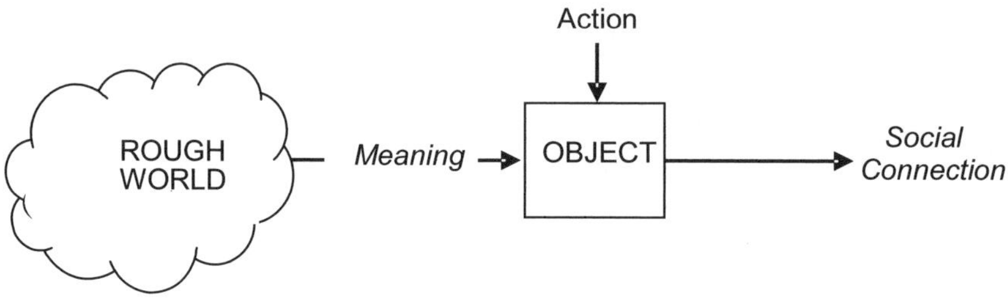

Figure 7.1. *Emergence of the object*

To explain this schema, here is a quote from a book entitled *La Ville Émergente* by Le Guirriec [LEG 97, p. 80]: "I go two, or even three, times a week to Leclerc in Brest, but I buy my clothes at Casino. I do not shop at this grocery store because I don't want to pay their prices. I prefer to go to Bourg-Blanc where I go several times a day to such a point that people think I live there. I buy my bread at Bohans because I like the bakery. My doctor is in Milizac, but my dentist is in Bohans, because the one in Milizac has no patience with children. My son plays soccer at Coat Meal, but next year he will go to Bourg-Blanc where the level is higher."

This description of the relation of a Breton mother with her territorial environment shows an "instrumental proxemics" type of relationship, a relation that unities her with her family, its providers, its doctors and dentist, to form a sort of network.

If we go return to the context of territorial cooperation and to its project, we can group all our reflections in the diagram given in Figure 7.2.

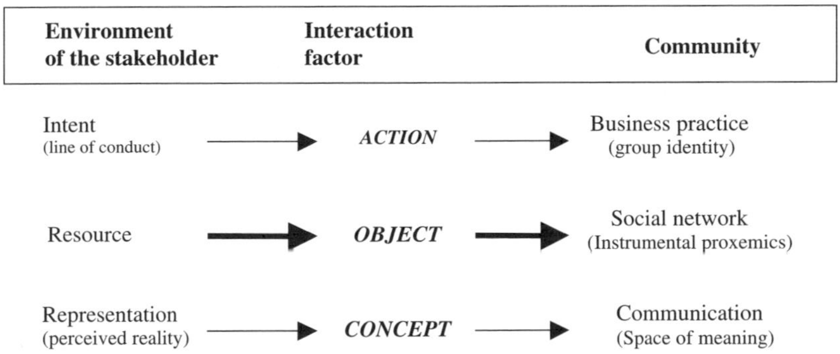

Figure 7.2. *From stakeholder to community*

Intent gives meaning to action [DEN 90], and this action updates the object to create a social connection with all the stakeholders directly involved or not with this object. The object then enables the creation of a space through what we have called "instrumental proxemics". Finally, to be communicated, the representation (internal to the stakeholder) of the territorial object must "cling" to a common social representation and take its meaning in communication.

The stakeholder gives meaning to his environment by projecting it in an inter-subjective dimension belonging to the community. Through ACTION, the stakeholder carries an identity built around a system of values and representations. Through OBJECT, he fits in a social network, ensuring a sort of legitimacy. Finally, through CONCEPT, he negotiates his credibility by expressing the terms of his territoriality in the space of communication.

Similarly, in the context of innovation, Callon and Latour [CAL 91] proposed the integration of the object itself into an extended network, grouping it in "a 'meta-organization' […] of humans and non humans interrelating together" [AMB 96, p. 134]. Through this network, translation phenomena occur, that is, "the transformation of a specific problem statement in the language of another specific statement".

In the territorial field, this notion of the role of intermediary played by the territorial object was brought up in the debate on swampy-zone land planning in the United States (as discussed in Chapter 1, the object is called the "boundary object") [HAR 98]. The object was placed in a context of "social debate" involving land planning challenges.

7.3.3. *Territory: an appropriate space*

In previous sections of this chapter, we presented territorial stakeholders in the context of territorial cooperation and their relation to the territory through territoriality. Territoriality, as it was defined, is based on cognitive stakeholder models and is built on the basis of an appropriation of the space that transforms it into territory.

For Brunet *et al.* [BRU 92], the territory is "an appropriate space with the feeling of conscience of its appropriation". In another context, the context of ethnography, Mauss [MAU 47] indicates that "a determined social group has its hunting territory where it will not escape the limits, even if it is nomadic. The native knows his territory: water supply, plants, nature, number and habits of animals; outside, he will often feel lost." In this last quote, we notice that the native has a network of information serving as indicators for the description of his territory according to its use.

Closer to our topic, Poche [POC 96] shows that the identity of a social group and the notion of territory are highly linked, to the point that we can say that the representations given to the territory represent a sort of language of the social group's spatial context.

Following these ideas, we believe that the relation of appropriation linking the stakeholder to the territory is first and foremost a relation of space qualification. Depending on the territory's use, its desired purposes, and then stakeholders' legitimacy, the territory will carry characteristics representative of this appropriation.

This is also highlighted by M. J. Chombard de Lauwe in [BAS 97, p. 148]: "Appropriating a location is not only recognized as being able to use it, it is establishing a relation with it, integrating it in everyday life, being able to mark it with our footprint and become the catalyst in its transformation".

In this way, the appropriation of a space as territory corresponds to two combined processes:

– *The definition of a range*, making a controlling space a reality, even if the limits of this range remain fuzzy because they depend on the capacity of the stakeholder to establish himself in the play of multiple interactions occurring in the same space. This uncertainty is involved in the creation of "informational slack" elements.

– *The activation of cognitive structures* of the person, social group, or institution (which we have called the "stakeholder") appropriating the space. By cognitive structures, we mean processes of knowledge, representation, and understanding of the world. These processes are based on perception and learning capabilities and enable a cognitive interpretation of the environment.

In this model, we find, in Figure 7.3, the following:

– The "territory" environment, a resource space rooted in the natural environment (on the right), that is subject to the skill the stakeholder believes he has to control the natural environment (the object of the relation (1)) for specific purposes.

– The cognitive environment associated with this concrete "territory," as it is interpreted by the stakeholder in his cognitive understanding (object of relation (2)) with the help of the cognitive models that he implements. We can explain in a lexical analysis the elements of these models.

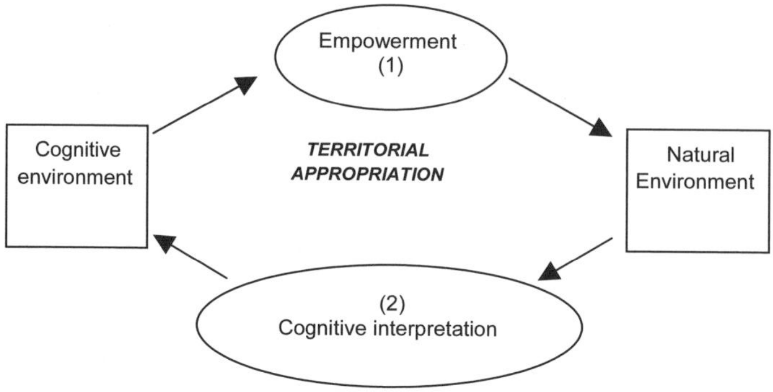

Figure 7.3. *Territorial appropriation*

This model associates the natural environment with a cognitive environment, giving it meaning; on one hand, action on the territory is possible because of the feeling of empowerment that the stakeholder has, and, on the other hand, cognitive interpretation of territorial elements relies on the "reality" of the natural environment to build a cognitive environment, thus making it possible to distinguish the interesting territorial objects. This cognitive interpretation occurs because of the activation of the cognitive models of the territorial stakeholder. These cognitive models are developed by the practice of associating action in the territory with acquired knowledge.

In other words, "stakeholder" and "territory" cannot be dissociated because we must take into account the stakeholder's purpose, rationality, cognitive environment, intentions, and empowerment in the context of his involvement in the territory [MAJ 96]. Territorial cooperation revives the territorial appropriation game and forces the expression of territoriality.

7.3.4. *Cognitive environment and territoriality*

The cognitive environment presented in the previous model (Figure 7.3) is the environment of territorial representation as it emerges from territorial practice and cognitive interpretation. It can be defined as a mental space where significant elements from the territory for the stakeholder are identified.

The cognitive environment can be considered an environment of implicit knowledge arising from the expression of territoriality. That is why the territory has a "natural" legibility. The stakeholder builds a mental image of the territory that can be described and, especially, communicated. In the following quote, Lynch [LYN 98, p. 9] qualifies the mental image of the territory by emphasizing the uniqueness of this logic of construction:

> *"An image of the environment can be analyzed through three components: identity, structure and significance. [...] A usable image first requires the identification of an object, which presumes that we can distinguish it from other objects, that we can recognize it as a separate entity. This is called identity, not in the sense of equality with something else, but in the sense of individuality or uniqueness. Second, the image must understand the spatial or paradigmatic relation of the object with the observer and with other objects. Finally, this object must have significance, be practical and emotional for the observer. Significance is also a relation, but completely different from the spatial or paradigmatic relation."*

According to the authors of the chapter, the meaning, which guarantees the consistency of the cognitive environment and qualifies the space into territory, is territoriality. This meaning is developed by social practices organized around territorial objects; these practices define, or even reinforce, relations within a group. These practices use elements from cognitive models acquired by the territorial stakeholder. They are realized in what we have called "instrumental proxemics" in terms of the physical dimension of territoriality, and the representations and words of the language involving notably the cognitive dimension, and finally in the definition of rules and symbols marking the territorial structure of space involved in the symbolic dimension.

If we refer to GIS, we realize that, in this context, GISs are only involved in one dimension of territoriality, that is the cognitive dimension. This explains the reservations we have and will present at the end of the chapter on the limits of current GISs.

7.4. Cognitive model diversity of territorial stakeholders (e.g. Switzerland)

We will focus here on the analysis of the diversity of the cognitive models of four institutional stakeholders: the developer, working on territorial land management; the urban planner, practicing in the urban area; the transport engineer, organizing circulation methods; and the heritage preserver, looking after valuable historical objects.

7.4.1. *Lexical analysis of stakeholder views: methodology and results*

We carried out a lexical analysis of the views of these stakeholders [MAJ 99], an analysis in which Serrate [SER 95, p.127 to 129] showed the relevance of a territorial context on the basis of samples representative of their documentary production. The objective of the analysis is to describe the elements of these stakeholders' cognitive models by using the definition of territoriality established from its three dimensions (physical, cognitive, and symbolic) and by considering the elements required for territorial cooperation.

Territorial stakeholders have a job. By "job" we mean a practice developing its own logic and defining a professional identity in the social structure in which it is involved [SAI 77]. The notion of "job" also has a close relationship with the notion of knowledge, meaning by that that all knowledge is mobilized in a work practice [OST 03].

The territory is one of the fields in which a work practice is vital to the achievement of projects. In the context of the practice of these trades, there is unity between decisions made and actions carried out.

In terms of the principles of interpretation of a lexical analysis of speech, we can refer to Gavard-Perret and Moscarola [GAV 95], who present the characteristics of the statistical study in great detail. This type of analysis is based on the creation of a crossed lexical table, in which we conduct a factorial analysis of correspondences (FAC) to bring out proximities between units of significance [MOS 94]. This analysis brings out the "key" words belonging to the cognitive model of the author of the views. Finally, the FAC requires a thematic interpretation according to the nature of the grouped words; that is why we have defined thematic dictionaries to add consistency of interpretation and avoid any insignificance.

For this analysis, we used the "Sphinx" software specialized for data and lexical analyses.

This lexical analysis brought out the following results.

Cognitive models of the four trades are clearly identified. We can say that the developer ["LAND MANAGEMENT"] is centered on large zones of the territory by speaking of region and development. Conversely, the urban planner ["URBAN PLANNING"] limits his field of action to zones and districts. The heritage preserver ["HERITAGE"] focuses on a context (fabric, constructed, morphology) while remaining in the cultural field. The transport engineer ["TRANSPORT"] has a linear vision to for moving and exchange. In general terms, the FAC highlights the fact that the elements common to the different trades are the elements chiefly responsible for the project's dynamic.

Space and time are also factors of trade differentiation. Some trades have a wide vision of space (land management, transport), whereas others focus on delimited, clear zones (urban planning, heritage). Separation of trades is clear in terms of the vision of time. Heritage is the only trade operating through a projection of the past toward the future; the others project into the future. We can speak of multicolored vision in this context [HAL 83] because this groups the different eras of the project by bringing them into the future. We also note the proximity of approach between land management and transport with a monochrome vision [HAL 83], that is, one separated into individual pieces even if the period is different. The urban planner seems to be in a situation of event repetition.

Urban planning and transportation businesses are in the field, whereas land management and heritage are more conceptual. We should also note that land management searches for good territory organization, that urban planning studies the urban environment by sector and zone according to plans (good scale complemented with land management), that heritage moves toward cultural elements, and, finally, that transport cooperates with urban planning to create the city.

In terms of the symbolic dimension of territoriality, heritage questions value and significance, urban planning questions quality and the consequences of densification, land management questions the large territorial balances by considering the challenges to be solved, and transport strives to improve the situation in terms of accessibility and mobility.

In summary, the terms of the underlying cognitive model are radically different, some centering on the reality in the field (transport and urban planning), and others on concepts (heritage and land management). The conflicts can involve the details of the project (transport, urban planning), which will require strong coordination or prior evaluation (land planning, heritage), and which will require arbitration. The different visions of the territorial stakeholders also explain the problems encountered in implementing an efficient coordination for a common project. Arbitration is one of the possible ways to respond to this situation.

7.4.2. *Symbolic history of territoriality and GIS*

The symbolic history explains the argumentative strategy of the stakeholder and expresses his vision of territoriality, in accordance with his purpose of intervention in the territory. This history can somewhat summarize his knowledge in action.

For us, the symbolic history, as a support to our argument, follows a "ternary narrative" [RIC 90] to "describe, say, prescribe". Ricoeur [RIC 98] defines three aspects of this narrative:

– "prefiguration," corresponding to "the realization of togetherness";

– "configuration," corresponding to the step where "the act of saying is freed from the context of daily life and penetrates the sphere of literature";

– "refiguration," corresponding to the act of "discovery, reveal the hidden…" for the narration.

We find these three dimensions qualifying territoriality: the physical because of its "instrumental proxemics" (*togetherness*), cognitive because of representation and language (*say*), symbolic because of the symbolic objects in the territory (*reveal the hidden*). The symbolic history is suitable for "describing, saying, prescribing" the relations of the stakeholder with the territory (i.e. expressing territoriality) and support its actions as well as its arguments in the context of territorial cooperation. Symbolic history is instrumental to action.

In more specific terms, symbolic history is used for:

– the possibility of the stakeholder building "intrigue" into the play of interactions with the other stakeholders;

– adopting a line of direction making it possible to plan his actions and coordinate them according to the results obtained;

– proposing solutions compliant with this vision;

– implementing an implicit knowledge base made up of a series of simple interpretations constituting the main aspects of the cognitive model;

– optimizing the efficiency of activities by integrating the functions of the "business" information system into the context of the action system.

A symbolic history can be defined through the interpretation of the FAC maps presented. For the four studied trades, we can obtain symbolic histories, as represented in Table 7.1.

Job	*Symbolic history/narrative logic*
Land management	*d: the territory organized in administrative division* *n: development problems* *p: finding solutions*
Urban planning	*d: construction integrated in a district concept* *n: realities of the field* *p: propose the best urban planning possible*
Transport	*d: transport system in the city and suburb* *n: accessibility* *p: set urban movement*
Heritage	*d: qualified space-time* *n: architectural heritage* *p: question on value and sense of territorial objects*

d = describe; n = narrate; p = prescribe.

Table 7.1. *Symbolic histories of trades studied*

If we refer to both main aspects of spatial reference information systems, "memory" aspects (Describe) and "decision support" aspects (Narrate), and to Système d'Informàtion à Référence Spatiale (spatial reference information system) terminology proposed by Thériault [THE 92], adapted from Bédard [BED 87], we make the following comments on this table:

– For LAND MANAGEMENT, the information system is Système d'Information Territorial (territorial information system) with a relation with territory that is much institutionalized in administrative divisions (communal limits, zones, etc.). The "decision support" aspect must describe the problems linked to the development of territorial activities. In terms of the result, the decision, we must find solutions to the development problem with other institutional stakeholders. A collaborative approach must support the territorial information base.

– For URBAN PLANNING, the information system is centered on the job of construction integrated into a district or zone concept. this somewhat confirms the reflection of certain current schools of thought in urban planning maintaining that currently we must consider the city as a mosaic of sectoral planning visions instead of a single consistent unit. The information system then is a geographic information system, with elements specific to the approach by integration (building outline and alignment, access processing, aspects of social life, etc.). The "decision support" aspect must account for this possible integration by the explanation of the realities of the field. Finally, the decision itself must involve proposition in term of urban planning; simulations or impact studies are therefore particularly relevant.

– For the TRANSPORT sector, the information system favors systemic aspects (using words such as "system" and "connection"). The transport system in the urban field represents the fields of intervention. Accessibility of the city sectors gives a motivation of action; the objective is to find a regulation for the movement flows. In our opinion, we are facing "operational search" problems, which can result in the use of complex mathematical models.

– Finally, for the HERITAGE sector, the context of time period is important. The information system must be able to refer to these periods to help in the determination of the value and sense of interesting objects. The objective is to argue according to period references. It is for us a "knowledge base" information system and an approach by means of elements located in a historic context.

In assembling a number of previous reflections made in this chapter, we have created Table 7.2, distinguishing the elements of knowledge (static) from the elements of the dynamic. This table shows the variety of architectures of information systems. Recognizing this variety necessary to really consider the diversity of territorial approaches of trades.

Job	Elements of knowledge	Elements of the dynamic
Land planning	Classifications, statistics, simulation models taking into consideration territorial zones, multi-criteria or sectoral models highlighting the great balances in terms of planning and space	Collaborative approach, "elaborate groupware," Système d'Information Territorial (territorial information system)
Urban planning	Descriptive representations, prototypes in the context of the environment, incidence of the density and lifestyle, multi-criteria impact models	Approach by integration, "decision support", GIS
Transport	Statistics, classification by zones, simulation models, mathematical flow determination models, multi-criteria models	Systemic approach, "operational search", Système d'Information en Recherche Opérationnelle (operational research information system).

Heritage	Descriptive representations, representation model in the context of the site, situation of interest, knowledge models, document bases	Approach by elements, "knowledge base", Système d'Information Documentaire (documentary information system)

Table 7.2. *Elements of knowledge and dynamic of trades studied*

7.5. Implications of cognitive model diversity on GIS

7.5.1. *Role and limits of current GISs*

To clarify the role of current GISs, we retain the three dimensions described by Lescat [LES 95] in defining an information system: the human dimension, for the skills necessary for controlling the information system; the technological and technical dimension, for data processing and formatting of information; the organizational dimension, for organizing tasks.

In GIS, we observe:

– the resulting development of "geomatic" skills, making it possible to simplify georeferencing;

– the emergence of new computer tools to treat the spatial references and to describe territorial objects according to predetermined thematics (cadastral registration, zoning, public equipment, communication lines, land improvement, rivers, etc.).

An observation we can make, however, is that GIS is not at the heart of territorial cooperation as a common federating tool.

Observations from previous sections help up identify several reasons for this:

– First, and most importantly, GIS has the stake in a balance of power between the partners, as shown by Pornon [POR 97]. Because they perceive the wealth of territorial information and the possible use of representations, the stakeholders tend to want to appropriate the GIS for their own profit and do not want to make it a common element of coordination.

– Then, even if the spatial reference enables each stakeholder to make his action in the territory, the territorial object always remains the main center of interest. The representation is never substituted to the object itself, the object around which the cooperation and the power struggles described earlier is actually organized. The representation is only used to make the necessary information available and to support the argument of the stakeholders involved.

– Finally, even if currently we continually focus on the pairings of different tools, as, for example between GIS and decision support tools [JOE 00], the GIS remains for the stakeholders mainly a method of storing information organized according to their concepts of territorial interpretation.

7.5.2. *Suggestions for implementing GISs*

In section 7.4, we established that the cognitive models of the varying trades are slightly different from each other. Three points vital for GIS arise out of these differences:

– 1. It is delusory to believe that a single information system tool can represent, and situate itself in the cognitive logic of trades, the data sets necessary to a territorial project.

– 2. In fact, GISs are poor in terms of dynamic coordination itself. There is still a lack of tools available to feed and instigate the debate necessary to the development of a territorial project; the tools proposed by Pornon [POR 97] (an integration of highly different tools) and Joerin and Musy [JOE 00] (decision support tools) are required and their acquisition are required and should be improved.

– 3. And yet, it appears necessary to develop information systems specific to the trades to be able to account for their territorial vision and their territoriality, with the reservations addressed in the previous point maintained.

Contrary to an idea often mentioned, GISs are not revealed to be federating tools. They could even be used as weapons for the stakeholders when negotiating their interventions in the territory. They allow the stakeholders to provide their best arguments, enabling them to consolidate their positions (by analogy to the models discussed by King [KIN 93]). A process of comparison of cognitive models from stakeholders can only be considered in the context of a territorial project involving negotiation and cooperation [GOL 95]. Chrisman [CHR 00] also agrees with this perception, warning us against the current tendency to build too rigid of ontologies, not giving enough room for the ambiguities that are vital to human action.

We favor the development of GIS in respect to the following:

– 1. GISs that satisfy the inherent needs of each job must be developed first, in accordance with the elements of the dynamic as established in Table 7.2.

– 2. GISs must facilitate the decision processes by providing the necessary arguments to the stakeholders involved to support their positions and by highlighting consensual, or at least mutually acceptable, solutions; the tools proposed by Jankowski *et al.* [JAN 97] and more recently by Sidlar and Rinner [SID 07] seem particularly promising.

– 3. Existing ambiguities between the perceptions of the different stakeholders and trades are argument and negotiation spaces that must be respected during GIS design in organizations.

7.6. Conclusion

In conclusion, it is important to recall some key points in reference to the three major fields of investigation that we have studied in this chapter: territory and territoriality; cognitive models of decision-making stakeholders; and information systems. For each of these fields, we address once again what we consider their most important factors, returning to those sections corresponding to more detailed information.

Concerning territory and territoriality:

– 1. The territory can be defined as a space of resources as well as a space where the identity of stakeholders and their legitimacy in the cooperation game can be negotiated.

– 2. Territoriality is a meaning and not a property; this meaning emerges through the activation of social practices in the physical, cognitive, and symbolic dimensions of relation to the territory.

– 3. Territoriality can be described, in terms of is major significance, by a symbolic history that the trades deploy in their territorial practice.

In terms of cognitive models of decision-making stakeholders:

– 1. The cognitive models of the trades (LAND MANAGEMENT, URBAN DEVELOPMENT, TRANSPORT and HERITAGE) are completely different from each other.

– 2. Job specialization reinforces the trades' autonomy and requires increasingly frequent arbitrations.

– 3. The challenges of cooperation for the trades are as follows:

- LAND MANAGEMENT: consensus at all costs;
- URBAN PLANNING: legalism;
- TRANSPORT: technocracy;
- HERITAGE: cultural dogmatism.

In terms of information systems:

– 1. Because of the diversity of cognitive models and the varying natures of information systems, there cannot be a single, federating GIS.

– 2. Lexical analysis is revealed to be an efficient tool of study, a tool with the ability to highlight the cognitive spaces inherent to each job and the spaces that could be used as a connection between jobs.

– 3. Territorial objects have a natural polysemy that must be respected in GISs representing them.

– 4. Information systems associated with trades must enable a representation of territorial objects, but they must not be limited to this functionality. It is necessary to add other functions to a GIS to retain the specialization of trades while supporting a dialog through common references or integration processes.

7.7. Bibliography

[AMB 96] H. AMBLARD, P. BERNOUX, G. HERRERIS, Y.F. LIVIAN, *Les Nouvelles Approches Sociologiques des Organisations*, Paris, Le Seuil, 1996

[AUS 95] G. AUSLOOS, *La Compétence des Familles. Temps, Chaos, Processus. Relations*, Ramonville Saint-Agne, France, Érès, 1995

[BAS 97] M. BASSAND, *Métropolisation et Inégalités Sociales*, Lausanne, Presses Polytechniques et Universitaires Romandes, 1997

[BED 87] Y. BÉDARD, "Les différents types de systèmes d'information à référence spatiale," *Carto-Quebec Convention/Canadian Cartographic Association*, Quebec, 1987

[BRU 92] R. BRUNET, R. FERRAS, H. THERY, *Les Mots de la Géographie. Dictionnaire Critique*, Paris, La Documentation Française, 1992

[CAL 91] M. CALLON, B. LATOUR, *La Science Telle qu'Elle se Fait*, Paris, La Découverte, 1991

[CHR 00] N.R. CHRISMAN, "Building GIS without foundations: ontology from a social practice perspective", *Proceedings GIScience,* Savannah, US, October, 2000, 4–6.

[CLA 74] P. CLASTRES, *La Société contre l'Etat. Critique*, Paris, Les Editions de Minuit, 1974

[CRA 98] WJ CRAIG, *The Internet Aids Community Participation in the Planning Process*, groupware for urban planning proceedings, Lyon, France, 1998

[CRO 77] M. CROZIER, E. Friedberg, *L'acteur et le Système*, Paris, Le Seuil, 1977

[CYE 63] R.M. CYERT, J.G. March, *A Behavorial Theory of the Firm*, Englewood Cliffs, NJ, Prentice-Hall, 1963

[DEN 90] D. DENETT, *La Stratégie de l'Interprète. Le Sens Commun et l'Univers Quotidien*, Paris, Gallimard, 1990

[DOS 97] F. DOSSE, *L'Empire du Sens. L'humanisation des Sciences Humaines*, Paris, La Découverte, 1997

[FRI 93] E. FRIEDBERG, *Le Pouvoir et la Règle*, Paris, Le Seuil, 1993

[GAV 95] M.L. GAVARD-PERRET, J. Moscarola, De l'énoncé à l'énonciation: pour une relecture de l'analyse lexicale en marketing, research note, no 95-03, University of Savoy, 1995

[GOL 95] F. GOLAY, T.L. NYERGES, "Understanding collaborative use of GIS through social cognition: 'do you see what I see?'", in T.L. Nyerges, D.M. Mark, R. Laurini, M.J. Egenhofer. *Cognitive Aspects of Human-Computer Interaction for Geographic Information Systems*, Dordrecht, The Netherlands, Kluwer Academic Publishers, 1995

[HAL 83] E. HALL, *La Danse de la Vie. Temps Culturel, Temps Vécu*, Paris, Le Seuil, 1983

[HAR 98] F. HARVEY, N. CHRISMAN, "Boundary objects and the social construction of GIS technology," *Environment and Planning A*, vol. 30, 1683-1694, 1998

[JAN 97] P. JANKOWSKI, T.L. NYERGES, A. SMITH, T.J. MOORE, E. HORVATH, "Spatial group choice: a spatial decision support tool for collaborative spatial decision making," *International Journal of Geographical Information Systems*, no. 6, vol. 11, 577-602, 1997

[JOE 00] F. JOERIN, A. MUSY, "Land management with GIS and multicriteria analysis," *International Transactions in Operational Research*, vol. 7, 67-78, 2000

[KIN 93] J.L. KING, K.L. KRAEMER, "Models, facts, and the policy process: the political ecology of estimated truth," in M.F. GOODCHILD, B.O. PARKS, L.T. STEYAERT, *Environmental Modeling with GIS*, Oxford, Oxford University Press, 353-360, 1993

[LAN 95] M. LANDRY, "L'ambiguïté comme outil de gestion", *Revue Française de Gestion*, vol. 10, 125-143, 1995.

[LAW 89] P. LAWRENCE, J. LORSCH, *Adapter les Structures de l'Entreprise*, Ed d'Organisation, 1989

[LEG 97] P. LE GUIRRIEC, "Mobilité et identité dans la périphérie brestoise," in *La Ville Émergente*, L'Aube, 1997

[LES 95] E. LESCAT, *Gestion de l'Information et Performance de l'Entreprise*, Litec, 1995

[LYN 98] K. LYNCH, *L'image de la Cité*, Paris, Dunod, 1998

[MAJ 96] W. MAJOR, F. Golay, "Une approche systémique du territoire et de sa dynamique," *3rd European Convention on Systems Engineering*, Rome, 1996

[MAJ 99] W. MAJOR, Approche de la concertation territoriale par l'analyse systémique et l'analyse lexicale du discours des acteurs. Perspectives d'application aux systèmes d'information géographique, Thesis no. 2003, Ecole Polytechnique Fédérale de Lausanne, 1999

[MAU 47] M. MAUSS, *Manuel d'Ethnographie*, Paris, Payot, 1947

[MOS 94] J. MOSCAROLA, "Les actes de langage. Protocole d'enquêtes et analyse des données textuelles," *Ex-Machina Colloquium*, La Sorbonne, Paris, 1994

[OST 03] F. OSTY, *Le Désir de Métier. Engagement, Identité et Reconnaissance au Travail*, Rennes, France, Presses Universitaires, 2003

[POC 96] B. POCHE, *L'espace Fragmenté. Eléments pour une Analyse Sociologique de la Territorialité*, Paris, L'Harmattan, 1996

[POO 06] B.S. POORE, N.R. CHRISMAN, "Order from noise: toward a social theory of geographic information," *Annals of the Association of American Geographers*, vol. 96, 508-523, 2006

[POR 97] H. PORNON, Géomatique et organisations: contradictions et intégration des projets d'acteurs, Thesis no. 1684, Ecole Polytechnique Fédérale de Lausanne, 1997

[RAF 97] C. RAFFESTIN, "Une société de la friche ou une société en friche," *Collage*, vol 5, 34-55, 1997

[REY 97] J.D. REYNAUD, *Les Règles du Jeu. L'action Collective et la Régulation Sociale*, Paris, Armand Colin, 1997

[RIC 90] P. RICOEUR, *Soi-même comme un Autre*, Le Seuil, Paris, 1990

[RIC 98] P. RICOEUR, "Architecture et narrativité", *Revue "Urbanisme,"* no. 303, non volume, pp. 57-76, 1998

[SAI 77] R. SAINSAULIEU, *L'identité au Travail, les Effets Culturels de l'Organisation*, Paris, Presses de la Fondation Nationale des Sciences Politiques, 1977

[SCH 94] E. SCHWARZ, "A meta-model to interpret the emergence, evolution and functionning of viable natural systems," *Proceedings of the International Conference on Cybernetics and Systems*, Singapore, 1994

[SER 95] B. SERRATE, "Développement local et représentations sociales. Analyse lexicale des discours," *Revue d'Economie Régionale et Urbaine*, vol. 1, 125-140, 1995

[SID 07] C. SIDLAR, C. RINNER, "Analyzing the usability of an argumentation map as a participatory spatial decision support tool", *URISA Journal*, no. 1, vol. 19, 47-55, 2007

[THE 92] M. THÉRIAULT, "Les SIG: entre le mythe et l'utopie," *Journées d'Études SIG et Gestion des Sols: Potentialités et Limites d'Application*, Lausanne, 1992

[VAR 96] F. VARELA, *Invitation aux Sciences Cognitives*, Paris, Le Seuil, 1996

[VIG 91] G. VIGNAUX, *Les Sciences Cognitives,* Paris, La Découverte, 1991

PART 3

Examples of GIS Uses in Organizations

Chapter 8

GIS and Territorial Governance: Examples in Quebec and Tunisia

8.1. Introduction

The involvement of governments in the acquisition, management and use of geographical information is undeniable. What is not so obvious, however, is the impact on this involvement and on the information itself that the evolution of government–society–information relations play and will play in the current context of the increasing involvement of civil society in territorial decisions. A certain number of questions should be asked: what is the role of information in a territorial governance perspective? What are the consequences of such a use on geomatic technologies and data and on decision-making processes? Conversely, can we see the impact that these technologies will have on this "socio-informational" evolution?

This brief analysis would not pretend to be scientifically rigorous – on the contrary! It is based on the assessment of a few projects carried out in different countries with different problems and is simply an attempt to highlight trends.

Chapter written by Jean-Jacques CHEVALLIER.

8.2. Projects involved

Some projects in which we were involved at different levels in the last 10 years have enabled us to better understand the role and importance of geographical information (GI) in territorial governance processes. Before continuing with the analysis of the relationship between government organizations, GI and territory, this section will briefly summarize the main projects that have encouraged our reflection. These projects are listed in the chronological order (Table 8.1).

Country or region	Name of the project	Time frame of the project	Theme addressed	Interested parties
Quebec	GIR	1992–1997	Integrated management of resources in forestry	Quebec Provincial Ministries Forest industry
Tunisia	AMILCAR	1996–1999	Water and soil preservation	Tunisia agricultural ministry Rural technicians Farmers
Quebec	CBE	2001–2003	Water management by calibrated watershed	Basin council Municipal and provincial authorities Population and local organizations involved
Quebec	Territorial profile	2002–2003	Management of public lands	Quebec ministry of natural resources

Table 8.1. *Projects involved*

A brief description of the characteristics of these projects is given later. We should note that this description is not meant to be complete or to reflect the vision of the people responsible for the projects; it is simply a summary to be used as a reference later in this chapter.

A preliminary comment must also be made. When we speak of territorial management, and especially participative management,we cannot restrict ourselves to the activities of authorities or conversely only address non-governmental approaches. The territory and its resources are of vital public interest to all the populations involved, including involving the responsibility of governments and the support of all human activity, requiring that the interests of all individuals and social groups be taken into consideration. One of the challenges of the good use of GI for territorial management resides in its availability for non-governmental contributors, who very often do not have human, technical and financial resources to acquire, manage and use information that they would need. The projects involved later cover a wide array ranging from mainly government project (*territorial profile*) to completely private initiatives (*CBE*).

8.2.1. *GIR project (integrated management of resources in forestry)*

The goal of the project was to develop new ways of addressing the management of forest resources (mainly wildlife, water and ligneous material), by going beyond the generally adopted sectoral and ministerial approach to reach better integration of the different visions of the territory and its use.

The project involved several expert groups (wildlife, water, recreational tourism, forestry, socioeconomics, etc.), and their role was to propose new sectoral models. A coordinating group was responsible for assembling these modules into a coherent piece.

The desired miracle solution was not found, but we were able to develop a better understanding of the sectoral challenges and social and administrative mechanisms. Some of the results could be used in other projects, AMILCAR in particular, and in the CBE project.

8.2.2. *AMILCAR project*

This was an inter-University research project (financing: IDRC – Canadian International Development Research Center), in cooperation with the Tunisian Ministry of Agriculture. The AMILCAR project followed a similar prior project, but centered on the implementation of remote sensing and GIS technologies.

Planning for water and soil preservation must take into consideration, at each stage of the projects, a combination of biophysical and socio-economic factors. The current work methods and the technological methods available are not well adapted for this type of combination. The project's goal is to fill these gaps in the specific context of CRDA [Commissariat régional au développement agricole (Regional committee for agricultural development)] activities in Tunisia, with the development and practical implementation of a series of methods (methodologies, data and models and computer systems) adapted to the characteristics of the development and choice of development scenarios at local and regional levels.

The project made it possible to develop a methodological approach and computer tools adapted to cooperation in territorial development projects. Societal processes and the structure of the dynamic group involved were emphasized, and technologies (geomatics and multi-criteria analysis) were relegated to second place.

Tested in the context of a pilot project (water and soil preservation land management planning of a Tunisian governorate), these elements of the solution have highlighted two key factors: first, it is much more important to manage the cooperation process efficiently than to implement sophisticated technologies; second, it is much more useful to have lower quality information that responds explicitly to a request from the people involved in cooperation than to have rich and precise data that does not satisfy the actual needs.

8.2.3. *CBE project*

The *Conseil de bassin de la Rivière Etchemin* [CBE (Etchemin river basin council)] is a private organization, grouping all the parties involved in managing the water in the basin of this river located south of the city of Quebec. As this river is located in a region with rate of high-pollution from urban zones and from industries, particularly in the field of agrifood (pork production), the quality of its water is greatly deteriorated over the years. CBE wanted to assemble the groups involved to gradually correct this situation.

The cooperation that took place involved everyone, from the simple citizen and the mother up to provincial ministries concerned, including the farming communities and representatives of municipal authorities. The first phase, which is the project here, consisted of developing strategic directions, accepted by all parties involved, making it possible to target future actions.

The following are the two major dimensions of this project:

– the development of an environmental database, with the help of the Quebec Environment Ministry, based on the principles of "ecological reference setting" created by this ministry;

– a cooperation organized around a multi-criteria approach, combined with structured techniques for working in groups, helping in going from the formulation of individual concerns to the statement of scenarios for a strategic plan and the choice of a satisfactory scenario.

A development scenario was finally chosen, setting the major directions and priorities with respect to the choice of future actions. The main problems were in obtaining a good representation of all the environments involved, to ensure regular and continuous attendance of the parties and the structure and adoption by the workgroup of a constructive cooperation dynamic, instead of the attitude of coercion and dissent too often seen in this type of project.

8.2.4. *"Territorial profile" project*

The Quebec territory is made up of almost 80% of public lands; private land is mainly concentrated in the St. Lawrence plains. However, public lands support private individual or commercial activities, with provincial authorization or permits. Because of this, the provincial government is responsible for formulating the land-allocation plan, used as the basis for sectoral plans and the granting of permits or concessions or the signature of leasing contracts.

Planning for the use of public lands is done by administrative region (13 regions in the province). Because of the geographical diversity of the territory and its resources, the heterogenity of knowledge on this territory and professional and technical resources available by region, a pilot project was launched in 2001 for developing a prototype for territory thematic analysis tool. The general objective of the territorial profile was to provide synthesized and structured information on the territory, presented in a comprehensive and accessible form to be used as a foundation for planning and territorial-management activities [MRN 01].

The project in which we were involved consisted of evaluating the potential of such a tool from the perspective of its usefulness in the different steps in the process (current and potential) of development and management of regional development plans for the public territory.

This evaluation involved the needs relating to the development of plans, as well as the evaluation of requests from individuals or companies for specific uses; it also had to evaluate the potential for other uses for the tool proposed, its openness to other public or private uses. Finally, the potential impact of the development of such a tool on decision-making methods and processes had to be evaluated.

8.3. Characteristics of the territorial decision

Before addressing the question of information for the territorial decision, we should review the major characteristics of such a decision. First, the term *territorial decision* will include any decision with a geographical component, whether it is input (geographical factors with an impact on the decision) or output (decision with an impact on the territory, as a construction project or urban planning).

In general, a territorial decision is complex at different levels; without going into too much detail, here is a partial list:

– thematic complexity: a territorial decision rarely, if ever, involves a single aspect of the territory; all the dimensions (social, environmental, economical, sanitation, etc.) to consider require the consideration of the many factors involving a large number of scientific disciplines;

– spatial complexity: all the factors (input and output) describing the different thematics considered require a large volume of spatial reference data; its geometry can be diverse and complex;

– social political complexity: a territorial decision can involve several levels of government, social groups and individuals, with very different and potentially conflicting concerns;

– administrative complexity: the spatial character of a territorial decision means that it can involve several departments with very different or even divergent concerns in a single level of government;

– the complexity of the decision process itself: having to consider these different dimensions makes the preparation and decision-making process more complex and involves adopting structured (to control the situation as best as possible) and flexible (to adapt to the countless specific cases) methods.

We cannot detail here the analysis of these characteristics. However, we should keep them in mind while we address the question of GI required for territorial decision.

8.4. Territorial information, of course, but why?

It has become common to say that territorial information is a must for any government. We also have said that 80% of the information generated by a level of government was linked in one way or another to a geographical position. We also often hear that "a GIS is a decision support system", especially from software vendors, etc. However, the different projects mentioned here have made us aware of the problem that we can encounter in obtaining the appropriate information for a decision. A more detailed analysis shows that it is necessary to really understand the role of information in the territorial decision to define how we could better respond to the needs.

In particular, in the perspective of the use of information in the context of territorial governance and involvement of civil society in important decisions concerning the territory, it is therefore necessary to identify clearly:

– government activities relating to the territory;

– among these, those potentially involving a participative methodology;

– the conditions in which this type of participation occur.

On this basis, it will be possible to better identify the information needs.

8.4.1. *Governments and civil society*

Considered globally, we can say that the government responsibilities relevant to this problem are addressed here:

– managing the territory, its resources and infrastructures,

– offering a range of services to the citizen, which includes in particular the availability of necessary information,

– acquiring and managing information required for its activities and to respond to the needs of the population.

This type of decision-making activities is common in the range of the government's territorial responsibilities, whether it involves planning or direct intervention on private or public lands. And most of them could benefit from a cooperative approach, whether it is restricted to government instances concerned or open to wider involvement from the stakeholders in civil society.

This category includes everything that has to do with planning of land usage; depending on the country, we will speak of zoning, land management, urban planning, etc. The objective with this type of decision is to set a certain number of principles and rules determining *who, what and where:* who can carry out which actions or have what activities in what parts of the territory. The potential confrontation of individual and collective interests is obvious: although it is the government's responsibility to preserve the national heritage as much as possible, especially from an environmental point of view, it is also necessary to preserve, whenever possible, the usage potential of the capital invested in real estate (outside of collectivist systems, obviously). A fundamental condition of sustainable development is to reconcile the two extremes. This is translated by a certain number of restriction measures to property.

In some countries, the use of territory can be subjected to different legal entities depending on the type of land tenure. In Quebec, the portion of territory under civil law, corresponding to a private property regime, only represents approximately 20% of the total surface. The rest is made up of public lands in which the government can grant rights for different categories of private or commercial activities: hunting and fishing, tourism and recreation, forest operation or mining, etc. The consistency of these rights should normally come from the principles described and spatialized in land-allocation plans. In practice, we notice that the statement of these principles and the preparation of these allocation plans can create tensions or even conflicts with the different government agencies involved in the development of some industries. This is the case in Quebec with forest operation and the debate concerning its control: this economical sector is very important and the imbalance caused by power struggles can distort the debate and could have pernicious effects on the choices needed in terms of public land development. In a perspective of governance, this statement of fact must be taken into consideration in the studies leading to a territorial development plan.

We must also consider that the role of government does not stop with planning. In fact, a well-conceived zoning or allocation plan can prevent many conflicts. However, the implementation of plans, although developed with the greatest care, inevitably leads to situations where previously accepted compromises on the major principles will eventually resurface in the form of complaints, requests for deviation or simply during granting of permits in cases unforeseen during planning! Everyone is all for virtue, but not everyone is ready to become virtuous. Cooperation for the harmonization of territory usage will therefore have to go beyond the prevention of conflicts and also intervene in other forms to facilitate management and the resolution of local conflicts. These three stages of governance, prevention, management and conflict resolution, will be discussed later in the section relating to decision-making processes to consider in the development of GI databases.

Finally, we must highlight the fact that this concept of "involvement" does not necessarily imply openness from governmental decisions to civil society. The simple fact of inviting representatives of other governmental agencies to participate in the preparation of the decisionthat, in principle, belongs to one specific authority is also a form of participation. From the methodological point of view, these two types of methodologies seem very similar to us, and following theoretical reflections should be able to apply to all situations.

8.4.2. *Possible and desirable participation*

We should first clarify what we mean by "participation". The concept of participation will have, in fact, very different connotations depending on the country, legislations or cultures, and this catch-all term can cover very different realities. Figure 8.1 illustrates this diversity of explanations.

At one end of the spectrum, there is total control of the situation by citizens. This extreme situation is not seen in a stable legal context, because it would exclude any form of rule or government intrusion. The next stage of "delegation of power" is encountered in some cases, in particular when social political conditions are so that it would seem preferable to delegate the responsibilities of decision making to the civil society of a specific sector or region. These first two stages do not really represent participation, in the sense of a joint exercise of some powers, but more a transfer of power to a given social group.

The three next levels really imply participation, but may be tricky to implement for many reasons: fear of loss of government control, loss of power by civil servants (!), lack of skills, quasiabsence of adapted work methods, problems with practical implementation, etc. These problems are particularly obvious in the case of certain *issue tables*, created with commendable intentions to respond to real needs, but too often go around in circles. However, in [BUC 99], we find some successful examples of participation in environmental management.

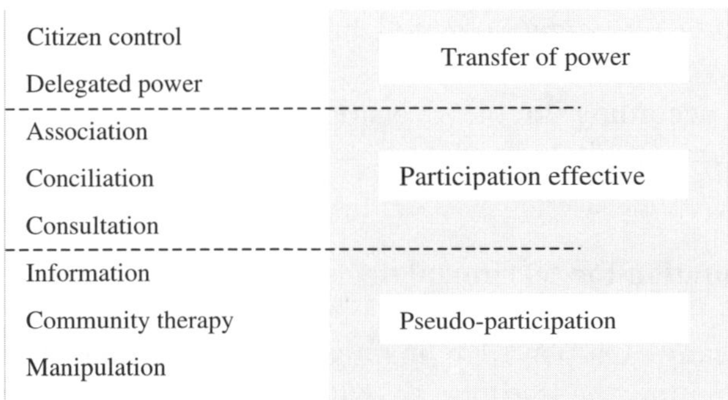

Figure 8.1. *Levels of participation (according to [ARN 69])*

In practice, the participation advocated by governments and promised by politicians is too often just simple information, often coming too late for any constructive reaction from populations. The last two levels of Figure 8.1 are self-explanatory.

We must moderate this comment, which may sound like criticism: in fact, active participation is not always obvious everywhere and to everyone! It causes fundamental paradigm shifts:

– in psychological terms, individuals must be able to feel concerned and be conscious that they can express themselves and participate in a constructive way;

– in social terms, we must consider a redistribution of the respective roles of government and citizens, implying that governmental organizations accept the fact that they do not control reality and that citizens have an increased responsibility for the environment and the territory;

– in decision-making terms, the mechanisms, methods and techniques of preparation and decision making must be reanalyzed according to the psychological and social evolution;

– in informational terms, we must grant each party access to information necessary to play their role.

In the specific context of this book, this last aspect of the *informational paradigm shift* will be the subject of a more detailed analysis. We will see how processes of providing information are organized according to the decision-making mechanisms to better understand the resulting needs.

8.5. Information for territorial decision making

Historically, GI played a mainly descriptive and cartographical role: representing the territory, so that a user could perceive it with the help of maps and other documents. Considering the information as a component of decision making naturally leads to reconsider this role and to analyze it under different angles: the decision-making *process* as such, and in particular territorial decision making, the different *terms and concepts* used to identify the decision-making GI, and finally the *functions* that the information must provide in a decision-making process and the impact of these functions on the production of information.

8.5.1. *Management cycle and decision cycle*

Speaking of decision making for the territory implies the joint consideration of two *cycles* close to one another by their structure, but different in the context in which they are considered: the *territorial management cycle* and the *decision cycle.*

We mentioned earlier that one of the main responsibilities of a government is to manage the territory under its jurisdiction. Territorial management can be succinctly described according to four main phases: diagnosis, preparation of the decision, implementation and follow up. This process is cyclical as illustrated in Figure 8.2.

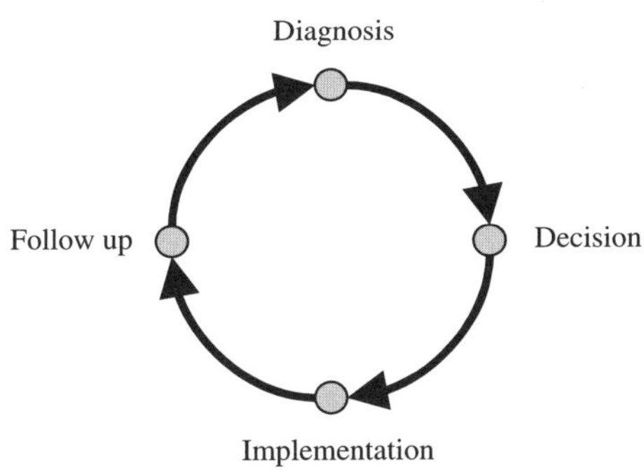

Figure 8.2. *Territorial management cycle*

The *diagnosis* phase must allow for the analysis of the territory involved in all its aspects: the needs (to identify the objectives of a possible intervention), potential (for understanding the possibilities of actions) and constraints (delimiting the operating area). Insofar as the territory is occupied or is the center of human activity, the contribution of the populations involved is vital, as long as the information available from authorities and government agencies cannot account for all aspects of the land, in particular its social and cultural dimensions. We find here a complement between *stakeholders in an institutional role* and *social agents*, as mentioned by [MAJ 99].

The *decision*, in the sense that we give it here, covers all preparation activities for the decision leading to the formulation of possible action scenarios to the formal decision.

The *implementation* is the concrete side of the decision taken in the previous phase and constitutes the territorial intervention itself.

Finally, any responsible management should include *follow-up* activities in the territory for collecting descriptive information from the intervention carried out during the implementation phase, and to observe and note observations relating to the continuous evolution present in any territory: natural evolution, impact of human activity, etc.

In the field of decision support, a model that cannot be ignored is the Simon model, published in 1960. However, we prefer a spin-off model proposed by [MAR 88] and [CHE 96] and presented in Figure 8.3. Although it is more detailed, this process corresponds precisely to the cycle of territorial management described earlier. *Intelligence* refers to the understanding of the object of study, *analysis,* to the formulation of solution scenarios, *choice,* to the identification of the scenario(s) to use, *decision* is the authority required to designate the scenario to implement and *accomplishment* is the result of the actual decision process. We completed the diagram with a retroaction loop, the *follow-up,* required to update the databases used.

It is important to be aware of the diversity of activities involved in these different steps, in particular what we are concerned with here, the role required by territorial information to accomplish these actions. However, before analyzing this diversity, we should somewhat clarify the terminology commonly used.

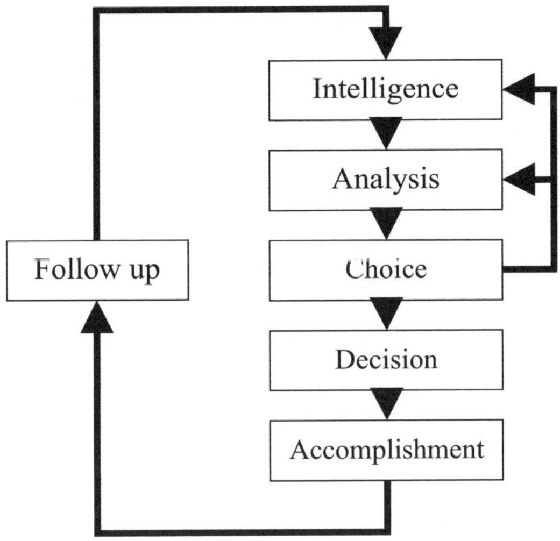

Figure 8.3. *Decision-making process according to [MAR 88] and [CHE 96]*

8.5.2. *Terminology used*

We commonly use different terms relating to information for the territorial decision, particularly the environmental decision. There is a great confusion in this terminology, since the same term could, according to the context, designate different concepts, whereas in other cases, the same concept will be designated by identical expressions.

The most commonly used terms are *indicators*, *indices* and *criteria*. We will not analyze in detail these different definitions here. It seems more relevant to clarify the different levels of concepts required for a good understanding of the *nature* of the *information* for the decision and mechanisms required for its production.

On this last point, two approaches must be considered: first, we can consider the available data, the definitions of recognized indicator[1] and the scientific models recognized to analyze one factor or another; second, we can examine the origin of the need for information and the role it must play, to find the best way to respond to this need.

8.5.3. *The different functions of information*

Without going into too much detail, we can highlight the diversity of roles that information can play depending on the decision phase involved (see Figure 8.3):

– at the phase of intelligence of the problem, information can play a role in identifying the problem and/or potential problems offered by the territory;

– during the phase of analysis, it is necessary to formulate one or more solution scenarios and to describe their significant elements;

– during choice, we have data necessary to identify, describe and measure the discriminating factors helping us in choosing between the scenarios developed;

– for the final decision, we must provide the user with a set of information synthesizing the methodology followed, solutions considered, the choice and its justification, all information generated or obtained during the previous steps, but which must be reformulated according to this specific use;

– the follow-up consists of collecting the data describing the changes brought or observed in the territory to update databases and to identify the needs in revising the decisions taken before and launch a new decision cycle;

– finally, the practice of territorial management highlights the permanent need for communication between the parties involved in

1 For example, criteria and indicators developed in the different organizations such as the *Workgroup on criteria and indicators for preservation and sustainable development of temperate and boreal forests*, better known as *Process of Montreal*.

this management; particularly since one of the main sources of problems resides in the lack of knowledge that all the people involved in problem solving and in the interests of others have. Sustainable management mandates the improvement of communication and awareness, to facilitate this reciprocal knowledge. This requires clear mechanisms and specific information.[2]

Data contained in government GISs are provided by modern technologies (remote sensing, photogrammetry, GPS, etc.), as well as by digitizing maps and traditional files.

Without addressing the tricky question of the integration of heterogenous geographical data, we should highlight the fact that, in a large measure, this data are destined for the human eye and is in the form of maps or images. This data feeds a large number of simulation software packages even if it is not always adapted to the specific needs of one model or another.

The diversity of information use leads us to examine a little more closely the process that should result in the production of information adapted (in form and in content) to the desirable use. The major aspects of this process are summarized in Figure 8.4.

An example from the "territorial profile" project will illustrate this observation. In this context, the ministerial workgroup identified and described a certain number of indicators from the territorial diagnosis, and among them, an indicator called *density of the recreational pathways system* (Table 8.2).

2 The field of cooperative learning is involved here. These methods exceed by far the scope of this book, but they have a significant impact on the need for territorial information.

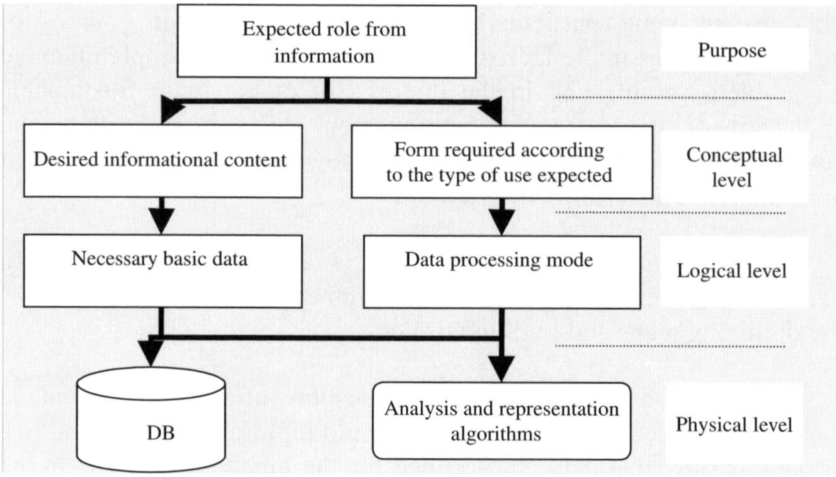

Figure 8.4. *Information for decision: from need to production*

Concept	Example
Role of information	Evaluate the potential of recreational tourism infrastructures
Informational content	Average density of the pathway system per km^2 of public territory
Required form	Thematic map (representation of the different densities through a color scale)
Basic data	Data on pathways from the Natural Resource Ministry (MRN) – Territory
Method of processing	Calculation of lengths of pathways and surfaces of analysis zones
DB	Système d'Information pour la Gestion des Territoires (SIGT) (Territory Management Information System)
Algorithms	Auto-Clip for calculating pathway distance Calculate Density of Spatial Analyst to determine density ArcView for cartographic representation

Table 8.2. *Examples of targeted production of decision-making information*

8.6. Effects on government information systems

The different aspects addressed in the previous section have a direct impact on the components of government information systems. We analyze them here in the context of resources constituting an information system: data, technologies, organization and human resources.

8.6.1. *Consequences on required data*

The data currently available in government information systems have certain characteristics making their systematic use for decision support tricky or even unpredictable:

– as we have said earlier, the vast majority of this information currently available is mainly cartographical; its use for other purposes, particularly in digital processing, specialized analysis, etc. must be done with the greatest precautions;

– except from the information contained in large-scale topographical maps most often dedicated to general use, geographical data were collected for a specific use or project; there again, taking it out of its context will not produce the expected result;

– this specificity of the purpose of data sets, combined with their large diversity, would mandate that we know in detail all the characteristics; the quality of the data is often unknown or difficult to understand; this weakness of available metadata was identified by several organizations, but fixing this problem requires a lot of effort and will not be done for all the data available, greatly limiting its potential for interoperability;

– decisions concerning the territory are rarely isolated decisions, but are often seen in a series of decisions, from the most global and strategic to the most local and operational; this mandates that the architecture and system technologies for decision support be built according to the concepts of hierarchy and multiple representations, similar to the principle of the cartographical scale hierarchy;

– the consequence of the traditional cartographical paradigm is a static approach to databases; territorial decision support not only requires updated data but also the possibility of having historical data and temporal data sets, which is not often the case;

– finally, the complexity of decisions, multiplicity of people involved and the diversity of methods, models and algorithms used to prepare the decisions makes the understanding of all factors leading to a decision increasingly difficult; the organizations involved will probably have to document in a more rigorous way their decision processes in relation with the data used.

8.6.2. *Consequences on technologies*

Traditionally, we have associated GI with a specific type of technology: GIS. However, considering the information as a component of territorial decision making creates new technological needs.

The diversity of data required and its origin results in the fact that traditional GIS software are less usable and the complexity of government informational structures aggravates the situation even more. Current technological trends offer new tools to respond to the challenges incurred by this situation. The field of *decision support systems* (DSS) is wide and constantly in development; we can only touch on it briefly.

We could suggest a classification of technologies:

– data-driven systems;

– model-driven systems;

– data-mining systems;

– decision-technique systems;

– people-driven systems.

Here is a review of these different categories.

Data-driven systems offer the user the possibility of using large databases for his analyses. In the spatial information field, the development of GI infrastructures (I^2G) and the implementation of *data warehouse* architectures can satisfy the need of accessibility for multi-source and multi-thematic data, a need that is constant during most of the decision phases described in Section 8.5.1. Specific workgroup constraints, as well as those from decision processes, are not really taken into consideration in this type of solution.

Model-driven systems are essentially autonomous systems executing "what-if" type simulations and other specialized analyses [CHE 96]. It is the response proposed by the community of specialists involved for a given problem; all simulation or forecasting software systems are included in this category, which brings together the efforts from the many specialists involved in its development. This type of model is desirable in the diagnosis, analysis and choice phase of the decision process; however, we must admit that these models are most often developed according to the interest of researchers and not according to the user needs: does the model correspond to the specific characteristics of the situation analyzed? Is the necessary data available? Does the proposed solution satisfy the needs? Do the calculated parameters help in choosing between the solutions considered? All these questions too often receive negative responses. There is an interesting bibliography on these first two categories of systems in [MEN 03].

Data-mining systems bring out trends and correlations from large volumes of data. In the particular field of geospatial data, recent developments lead to SOLAP software (*spatial online analytical processing*) [BED 03] [RIV 01]. The potential for this type of technology is huge, and the user interface can easily be controlled making this technology a precious tool, albeit limited to intelligence and analysis phases.

Decision support has long been the subject of several research studies, and applying this type of method to territorial questions is not recent. However, because of the thematic, social, political and administrative complexity of the territorial decision (see section 8.3), it seems that a multi-criteria approach [ROY 85] is highly desirable. Several studies addressed the question of integrating the multi-criteria analysis [CAR 91, LAA 00, EAS 93]. Some GISs actually integrate decision-support functions; that is the case with IDRISI, which offers a module based on the American school of multi-criteria (for a list of the types of multi-criteria methods, see [SCH 85]). However, this type of integration of multi-criteria analysis functions is not always satisfying, in particular, when responding to the needs of the choice of process phase in Figure 8.3. In this case, the idea is not to add additional functions of a GIS (which would result in a vision centered on data and on GIS software), but to combine a series of software components, which could include multi-criteria analysis modules [CHE 01] in a decision process. These functions can then involve raw geographical data (as is done with IDRISI) and serve as a tool for the consideration of preferences of the parties involved in a choice of a solution for a given decision problem [ROU 96, CHE 01].

The projects leading to a territorial decision often involve the participation of workgroups with sometimes large number of people. In addition, the geographical distance of the people involved and the frequency of meetings necessary for certain projects can create organizational problems. To offset these problems, *Group decision-making* software was developed. These tools facilitate information sharing, cooperation and interaction between the participants, as well as in certain cases, the implementation of virtual meetings in which each person can interact with the other while remaining in their office. Several products are available on the market today.

From a technological point of view, the main problem currently resides in the disparity of technological solutions, and to the best of our knowledge, none integrates all desired aspects. In addition, partial solutions are only rarely compatible with each other or with the work mode desired by the people responsible for the methodology to follow.

8.6.3. *Consequences on organizations*

The search for better use of information for territorial decisions will have major consequences on government organizations, their structures and modes of operation. The need to access a wide range of geospatial data mandates that all administrative barriers limiting this access must be destroyed; it is also necessary to overcome psychological obstacles and the widely spread feeling of propriety of the people responsible for the departments who possess this data.

A government strategy for GI acquisition, management and operation is therefore necessary. An example of such a strategy is the *plan géomatique du gouvernement du Québec* (PGGQ) (geomatic plan of the government of Quebec) developed in the late 1980s, and dedicated to developing cooperation and data exchange within the Quebec provincial government machinery. We must highlight that this type of strategy constitutes the organizational section vital to the development of I^2G, and probably the most tricky and lengthy to accomplish because of its impact on individual and corporate behaviors, responsibilities and power.

8.6.4. *Consequences on human resources*

All the developments described so far will have a direct impact on the people involved, in geomatic activities (data acquisition–management–distribution) and in decision processes. Changes caused by the technological evolution must be understood and accepted by all who are involved. These people should not passively experience these changes; on the contrary, they should be the main stakeholders and get directly involved in their preparation and their manifestation.

In addition, this change will require that skills improve for a varied and numerous work force, to prepare them for their future role. This is done with the implementation of continuous education and the development of new types of skills to control the integration of all the methodological, technological and informational components required for a sound territorial decision.

8.7. GI for territorial decision: toward a new paradigm

There are many geospatial data analysis and management software providers; their marketing departments have increased their vocabulary over time and according to trends: initially designated as *digital mapping*, software has become *geographic information systems* to *decision support systems*. However, no vendor can satisfy the technological, methodological, organizational and human needs caused by the paradigm shift!

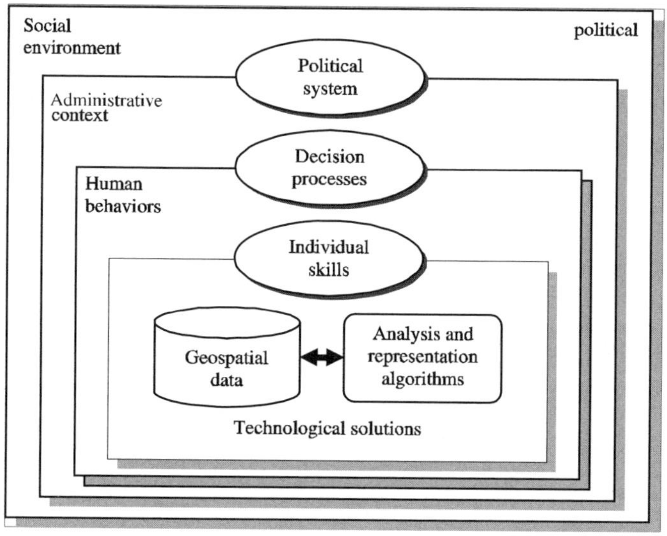

Figure 8.5. *The layers of the information system for the territorial decision*

The technological evolution triggered by the emergence of geospatial data in digital form reveals today an extraordinary potential for improving territorial decisions. The evolution of the *format* of the data is combined with fundamental change in the *role* of this information, the shift from analog map observed and analyzed by the decision maker to the digital use of data by sophisticated analysis software. At the same time, we can see an evolution in the future in terms of the type of data required and collected: from generic data, as the one used in the traditional administrative map, we evolve toward data sets satisfying the needs for use and models and specialized software; some organizations are even considering dropping a large volume of information they currently have because it no longer responded to the new methods of decision making and management for the territory.

The change must integrate in a harmonious way all the layers of the information system (in the broad sense of the word), as outlined in Figure 8.5. This evolution is fundamental from the point of view of the components of government information systems, and vital in terms of their capacity to play the fundamental role which is theirs in the future of our environment.

8.8. Bibliography

[ARN 69] SHERRY R. ARNSTEIN, "A Ladder of Citizen Participation" [archive], *Journal of the American Institute of Planners*, vol. 35, no 4, July 1969, pp. 216-224.

[BED 03] Y. BÉDARD, P. GOSSELIN, S. RIVEST, M.J. PROULX, M. NADEAU, G. LEBEL, M.F. GAGNON, "Integrating GIS components with knowledge discovery technology for environmental health decision support", *International Journal of Medical Informatics*, vol. 70, 79-94, 2003.

[BUC 99] D. BUCKLES (ed.), *Cultivating Peace – Conflict and Collaboration in Natural Resource Management*, International Development Research Institute, Ottawa and The World Bank Institute, Washington, DC, 1999.

[CAR 91] S.J. CARVER, "Integrating multi-criteria evaluation with geographical information systems", *International Journal of Geographical Information Systems*, vol. 5, 321-339, 1991.

[CHE 96] J.-J. CHEVALLIER, S. DAUDELIN, "La géomatique pour l'aide à l'aide à la décision en gestion des ressources naturelles: Exemple de la protection des paysages forestiers", *Revue Internationale de Géomatique*, vol. 6, 11-25, 1996.

[CHE 01] J.-J. CHEVALLIER, "L'analyse multicritère en tant qu'élément structurant des processus de concertation en matière environnementale", *Congrès FRANCORO III*, Québec, May 2001.

[EAS 93] J.-R. EASTMAN, P.A.K. KYEM, J. TOLEDANO, W. JIN, "GIS and decision making", in *United Nations Institute for Training and Research: Explorations in Geographic Information Systems Technology*, vol. 4, Geneva, 1993, 112 p.

[EAS 97] J.R. EASTMAN, *IDRISI for Windows. User's Guide, Version 2.0.* Clark University, USA, 1997.

[LAA 00] A. LAARIBI, *SIG et Analyse Multicritère*, Paris, Hermès, 2000.

[MAJ 99] W. MAJOR, Approche de la Concertation Territoriale par L'analyse Systémique et L'analyse Lexicale du Discours des Acteurs. Perspectives D'application aux Systèmes D'information Géographique, Doctoral thesis, École polytechnique fédérale de Lausanne, 1999.

[MAR 88] J.M. MARTEL, Aide multicritère à la décision, internal document, Département d'opérations et systèmes de décision, Faculté des sciences administratives, Laval University, 1988.

[MAR 98] N. MARTIN, B. ST-ONGE, J.P. WAAUB, "Geographic tools for efficient decisions in watershed management" in J.-C. THILL (ed.), *Spatial Multicriteria Decision Making and Analysis: A Geographic Information Sciences Approach,* Hants, GB, Ashgate, 1998, 309-334.

[MEN 03] A. MENDELZON, "Data warehousing and OLAP, a research-oriented bibliography", available at: http://www.cs.toronto.edu/~mendel/dwbib.html, University of Toronto, 2003.

[MRN 01] Ministère des ressources naturelles du Québec, Secteur du territoire, Direction générale de la gestion des terres publiques. SIGT-Territorial profile report (version 1.1 – préliminaire), December 2001.

[RIV 01] S. RIVEST, Y. BÉDARD, P. MARCHAND, "Toward better support for spatial decision making: defining the characteristics of spatial on-line analytical processing (SOLAP)", *Geomatica*, vol. 55, 539-555, 2001.

[ROU 96] H. ROUSSEAU, J.-M. MARTEL, "La décision participative: une démarche pour gérer efficacement les conflits environnementaux", Working paper 96-24, *CREDO*, Faculté des sciences de l'Administration, Laval University, 1996, 35 p.

[ROY 85] B. ROY, *Méthodologie Multicritère d'aide à la Décision,* Paris, Economica, 1985, 401 p.

[SCH 85] A. SCHAERLIG, *Décider sur Plusieurs Critères – Panorama de l'aide à la Décision Multicritère*, Presses Polytechniques Romandes, Coll. Diriger l'Entreprise, Lausanne, 1985.

[SIM 60] H.A. SIMON, *The New Science of Management Decision*, New York, Harper and Row, 1960.

[SIM 83] H.A. SIMON, *Administration et Processus de Décision*, Paris, Economica, 1983.

[VIS 99] Visual Decision Inc. "Decision lab 2000 Executive. Getting started guide", Montreal, 1999.

Chapter 9

Digital Participation and "Access" in UK Local Government

9.1. Introduction

This chapter has two *foci* relating to the emerging information society in Europe. The first is based on research into "digital participation" [SMIT 01], the role that Information and Communication Technologies (ICTs) can play in public participation within local authorities in the UK. The second focus relates to a wide series of issues that involve access to Geographic Information (GI) and the role it can have in participatory settings.

This research not only explored applications, technological developments and policy. It also attempted to understand actors' perceptions of public participation, drawing experiences from both officers who initiated exercises and citizens who chose to participate online. As such, this research uncovered some of the social and technical complexities of digital participation and what "participation" can mean and become for all actors. Part of this understanding involves the ways different methods can be adopted for

---

Chapter written by Robin SMITH.

public participation. This includes the emerging role that Geographical Information Systems have been playing, especially those that rely on the Internet as a foundation or conduit.

What many of these digital technologies offer is increased "access", in a number of senses, and the relationship this has with actors' understanding or notions of "public participation". One important group in this context are public sector bodies who produce and maintain information that is often used for participation and decision-making at various levels of government. Notably, their information is also one of the fundamental resources behind information economies that play a major part in European Information Society policy.

The chapter, therefore, draws on a number of sources to highlight these issues, ranging from European and UK policies for an information society, recent research and calls to formalize the activities in this field. Specifically, the discussion begins with an examination of the policy context for access to information and electronic government (e-government), where such policies were empirically tested through this research. A brief overview of the UK local government system and how Geographical Information Systems (GIS) have diffused into local government [MASS 96] offers a context for digital methods of participation. Following a discussion of recent activities in the UK, a theoretical model of participation's "components" is then introduced before adapting the framework to offer an approach to examine access to (geospatial) information.

9.2. The "Information Society" in European and UK policy

9.2.1. *European policy*

In the Member States of the Europe Union (EU), policies relating to access to electronic information stem from the Directorate General Information Society and the Information Society Technologies (IST) Programme. This programme funded the Geographical Information Network in Europe (GINIE: http://www.ec-gis.org/ginie) a group of

GI experts from industry, government and research supporting the development of a GI Strategy for the wider Europe from 2001-2004. GINIE reviewed many EU activities that apply to access to geospatial information and Craglia *et al.* [CRAG 03] described the policy evolution in this area within the context of geospatial information-sharing through Spatial Data Infrastructures (SDIs), which are closely related to the following discussion of wider EU information policies. In addition, SDIs share some similarities with digital participation as both rely on an understanding of social and organizational issues as well as technical ones, where this discussion hopes to provide links between the two.

The basis for developing Information Society polices was to increase the economic competitiveness of Member States in an information economy but not solely. Early policies tackled other issues relating to social exclusion and the digital divide [EURO 93]. Stemming from INFO 2000 (1996-1999) [DGXI 96], at the time of the research, eEurope policies are helping to set national and local government approaches to information access and digital participation. For example, they sought the availability of information as a "consumer right" [EURO 1a], a desire for everyone in Europe to have and use digital technologies and to have digitally literate and skilled citizens [EURO 1b]. More developed e-government has received support through deliberative online *fora* [EURO 1c], alongside increased accessibility to online services for disabled citizens [EURO 1d] and helping those in rural locations suffering from many barriers related to the "digital divide" [EURO 1e]. As such, although eEurope has an initial desire to develop information markets it also recognizes the importance of access and the need to support citizen participation in digital contexts.

Access to Public Sector Information (PSI), in particular, has been of interest to policy-makers since 1989 when a set of guidelines aimed at its promotion, increased transparency and an equality of access. Although widely discussed, the outcomes were largely ignored because of a lack of legal backing and an immature market [CRAG 03]. A decade later, the Green Paper on *Public Sector Information: a key resource for Europe* was produced, increasing interest in reused (digital) PSI and highlighting legal and financial barriers, before becoming a draft Directive in 2002 (Ibid.). Significantly, the draft does not address issues of access to any data, with local, regional and central governments supposed to find ways of doing this. In addition, most PSI benefits from GI, as it has a significant economic value and readily integrates policies from varying sectors (Ibid.). Values vary from €30-45 billion per annum (Ibid.) to £100 billion for the Gross Value Added from one National Mapping Agency alone (Ibid.), with differences being attributed to definitions of which businesses utilize GI. Integration is clearer to define though national (or regional) Spatial Data Infrastructures (SDIs). These frameworks of policies, institutional arrangements, technologies, data and people make it possible to share and make use of GI. Craglia *et al.* [CRAG 02] conclude from a review of European SDIs that political support at "global-to-local levels" is absolutely crucial for their development. This will clearly heavily involve local government as the producer and maintainer of much of this geospatially-related information.

Access to environmental information has been a particular feature of European policy. Many environmental lobby groups (such as Friends of the Earth) have demanded universal access to publicly-held environmental information to generate meaningful environmental policies, including sustainable development. Before the rapid expansion of the Internet, European member states had to abide by the 1990 Directive on *Access to Environmental Information* (90/313/EEC). Several years later, the 1998 *Aarhus Convention on Access to Information, Public Participation in Decision-making and Access to Justice in Environmental Matters* linked the right of access to information with human rights, including the right to participate in decision-making processes and to seek redress. This tied the concepts of "access", "participation" and social justice together. These policies, in turn, impacted on the development of the *EC Water Framework Directive* (2000/60/EC). This directive has meant that European river catchment information must be held on a GIS for data integration and analysis at the European level. A great deal of the information that contributes to this comes from local authorities and increases the need to examine local-to-global information-sharing. This Directive also provides the legal basis for the INfrastructure for SPatial InfoRmation in Europe (INSPIRE) initiative (http://www.ec-gis.org/inspire). Beginning at the end of 2001, INSPIRE aims to make relevant, harmonized and quality GI available for the formulation, implementation, monitoring and evaluation of European policy-making. To meet this aim, the initiative has been addressing a broad set of issues including common reference data and metadata (data about data), architecture and standards, legal aspects and data policy, funding and implementation structures and impact analysis. Its principal objective is to produce a European legal framework for environmental information matters, before providing a test-bed for other areas of Community interest, such as agriculture, regional policy and transport. Since the publication of the first edition, The INSPIRE Directive (2007/2/EC) has entered into force on the 15th May 2007. The Directive is currently in a two year transposition phase, with an action plan (2007-2009) developing metadata, interoperability of geospatial datasets and services, network services, data-sharing activities, monitoring and reporting of developments, the technical architecture and a geo-portal, that will provide access to the above infrastructure.

There is, therefore, a need to determine what local authorities' present abilities are to support access to information, both from within government and with outside actors, particularly in the e-government context. These policies show the importance of the formation of an "information society" at the European level, especially for recent developments for geospatial data; the ethos of which is reflected in the United Kingdom (UK) central government legislation and local government projects.

9.2.2. *UK e-government policy and research context*

It is useful to discuss the nature of UK local government as it provides the context for much the work in this chapter. Similarly there is a need to enrich this with how information society-related policy has developed in the UK, alongside how GI is used and valued in such organizations.

The United Kingdom is composed of Great Britain and Northern Ireland and, in turn, Great Britain contains England, Scotland and Wales. Over the period of the research there have been a number of reforms to the government structure such as Scotland's new parliament, a Welsh Assembly and ongoing work to establish a similar structure in Northern Ireland. Such diversity is also reflected in the UK local government system of 468 local authorities. Northern Ireland is a special situation and its 26 local authorities feature less in this research. Scotland (32) and Wales (22) have unitary authorities, where one organization delivers all local services but a more complicated situation occurs in England. Devolution has also occurred in London, in the form of an Assembly and directly elected Mayor, alongside the 33 London boroughs. Other cities have metropolitan borough councils (36) or English unitary authorities (47). All these have similar functions to the Scottish and Welsh situations. For other parts of England, a two-tier system of county (34) and district (238) councils delivers different services, with several districts dealing with local issues and services within a county, and counties dealing with more strategic policy matters. For example, at the time of the research, district councils dealt with local development control and local plans, county councils draw up strategic "structure plans" that inform local planning. Such a variation in government not only presents barriers to research but can also confuse citizens who may be trying to engage in public policy-making. It is within this context that e-government has developed in the UK.

The first UK research into the information society came from the decade long Programme for Information and Communication Technology [DUTT 99] that mainly had a technical focus [BRYD 98]. In terms of policy, the English Freedom of Information Act 2000 and concepts of "open government" provided early glimpses of how central government was coming to terms with the "information revolution" of the late nineties. The Act saw increasing local authorities' openness as empowering to citizens, providing rights to access information [CLAR 97]. However, as initiatives and legislation developed, various aspects of this policy lost their potency [SCOT 1a].

Scotland, having recently regained the power to make its own statutes via the new parliament, took a different approach. Access to Scottish PSI was seen as a right [SCOT 1b], reflecting the principles of the Aarhus Convention. Only "sensitive information" is exempt if a "substantial prejudice" coincided with its release [SCOT 1a]. Although beneficial to the Scots, variations in policy across the UK may prove problematic for information-sharing activities. Parallel to Freedom of Information legislation were changes in the nature of democracy, with a shift of emphasis towards understanding citizens' views. This was reflected in three policies: Information Age Government, Modern Government and Best Value.

Information Age Government was the core e-government policy in the UK that aims to have all government services deliverable through electronic means by 2005. A shift in political emphasis moved this target from an initial date of 2008. Although this was a major initiative, there is a lot of freedom in interpreting what "electronic means of service delivery" means, from advanced ICT systems to telephone call centre staff having access to digital records. A basis for this activity is "joined-up government", where a change in someone's circumstances, such as a "life-event" (for example moving house or having children), leads to a need to change government records. New systems are geared towards a citizen completing one form (possibly online) that will lead to all relevant governmental agencies being notified; thus, improving service to the citizen and increasing efficiency for government. However, there was some limited discussion away from service delivery to more democratic activities. Examples emerged such as some local government elections using electronic voting (to increase the numbers participating) and a series of more deliberative digital participation initiatives being mooted by central government [OFFI 02]. By 2005 targets were felt to be met but to varying degrees of service delivery and democratic engagement, arguably in part influenced by the performance indicators used to assess e-government success, noted below, and with new policies of transformational government providing a further focus on joining up processes, requiring increased access to information between public sector organizations.

At the same time as e-government was developing Modern Government was established as a means to change the political structures within local government, mirroring changes brought about by devolution. Various models were produced to help authorities shape their decision-making structures, such as a cabinet composed of ruling party members. Modern Government also encouraged local authorities to project themselves as civic leaders and a voice of their communities. Similarly, Best Value changed the way public bodies deliver services. It currently remains the framework for local authorities to continually improve relevant services that are democratically approved through consultation with citizens/users/customers and produces an abundance of consultation material. Authorities have to demonstrate their effectiveness through performance plans and reports which cover the value (financial or otherwise) of consultations, digital infrastructures and e-government services. A series of Best Value Performance Indicators (BVPIs) were introduced to help monitor improvements. Currently, as noted above, BVPI 157 reports on e-government performance for each local authority but such indicators are likely to altered in the near future, partly due to the adoption of new transformational government policy. It should be noted that many of the local authorities examined in the research were undergoing the earlier changes in policy context during the fieldwork period, as was the increasing value placed on GIS in e-government practice and policy.

Previous research, as part of the European Science Foundation GISDATA programme, examined GIS diffusion in local government. Work by Masser and Campbell [MASS 96] looked at the UK position, noting the extent to which (over 2 years), the adoption of desktop GIS had doubled and that mainframes were in decline. The research found that the technology had been introduced to different degrees in the types of authority noted above, with the south of England having more than Scotland. The same was not true in the emerging digital participation context. Although Masser and Campbell suggested that there was a need to maintain surveys of GIS in local government, little has been published, with a shift of research towards more technical applications, geospatial information policy and e-government. At the time of the research there was little evidence for GIS's role or an assessment of the impacts of e-government initiatives in general. Anecdotal evidence suggested that some GIS functions were moving away from technicians to officers (who are emerging from university with the training or gaining skills in the workplace) and at least one example of a web-based GIS has been used to support multi-agency decision-making with some public involvement (see below).

In addition, although providing a broad base, it has been noted that Information Society-related policies have not developed from "academic research from which critical theories of cyberdemocracy might evolve" [BRYD 98]. It was suggested that the lack of a critique has, in the past, resulted in limited empirical work and that the academic community has not recognized the need for "sustained attention" (ibid.). Therefore, the following discussion hopes to fill some of this gap by examining emerging examples of "digital participation" at a local government level in Europe.

9.3. Digital participation in UK local government

9.3.1. *Website survey and case study selection*

The investigation into local government-led digital participation was based around a survey of local authority websites in the Spring of 1999, followed by several case studies. This can be contrasted with the more "grassroots" activity discussed by Ghose in Chapter 10. The survey looked at over three-quarters of all ".gov.uk" websites at that time and attempted to find leading examples. The approach took on the role of a citizen trying to find information relating to their "community", local politics, the environment and land-use planning. As well as recording the different forms of communication that were promoted through the website, these organizations' "digital façades" were categorized in terms of the opportunities or facilities offered for online participation. The analysis drew on the frequently cited Arnstein's "ladder of citizen participation" [ARNS 69], a model that separates various forms of participation in a continuum from the manipulation of citizens to instances where the public control activities. Hunter recommends that the upper "rungs" of the ladder are particularly suitable for classifying local authority websites [HUNT 99] but initial analyses showed that the "top-down" nature of the local authority context only supported lower ones. The distribution of participation in the websites can be seen below (see Figure 9.1)

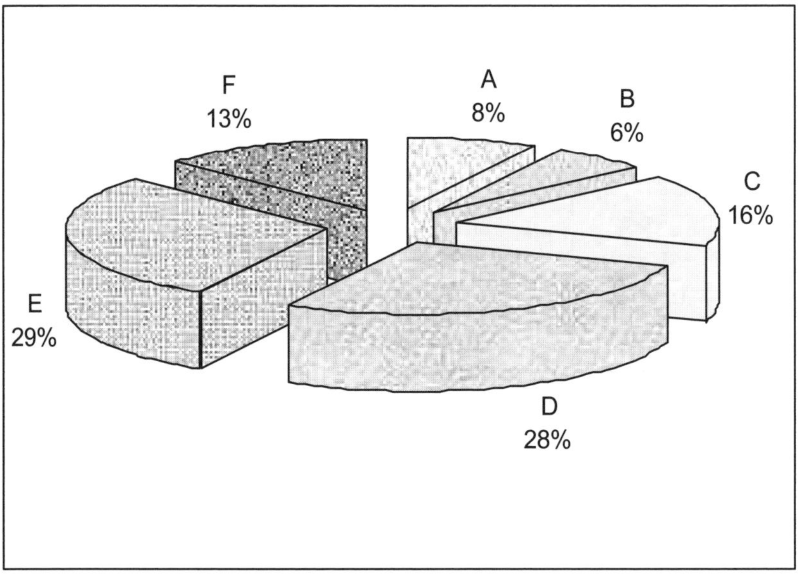

Figure 9.1. *The level of participation in UK local government websites, 1999*

Of the 287 websites examined, 8% could not be accessed (A), perhaps offering the greatest barrier to participation. Around 6% (B) had very limited content and were often "under construction". Beyond these poorer examples 16% of the websites appeared to be geared towards promotion through either tourism or economic development (C). The next category involved 28% of the websites and included those where very limited information about participatory activities was provided. This group, however, contained online government telephone directories and alphabetical lists of services, where citizens could then obtain further information. If there was more content, or if there were opportunities to interact online, then the "façade" and authority was seen as being an "information-provider" (E), presenting 29% of the websites examined. Finally, the "consultative" group contained 13% of the examples, equivalent to the middle of the "ladder" (F). This last group either contained full consultation documents online, online consultation exercises or more advanced digital methods such as chatrooms or bulletin boards. It is interesting to note that, unlike the previous research into GIS diffusion, leading examples existed across all forms of UK local government.

Although there was limited evidence of GIS being used through these websites, the survey provided a baseline for further investigation for both digital participation and instances where GIS was being applied to information-sharing, "access" or public participation. It also aided the selection of three case studies. These had four particular features: the type of participation the council was involved in; the methods they used; the issues being explored; and the different types of policy-making involved (local, strategic and mixed). The evidence for the cases was mainly collected from interviews with officers and citizens' and council documents. Overall, attempts were being made to look "behind" these façades, an approach informed by democratic theory and the social construction of technology (see Chapter 5). The context of each case is outlined next, before considering ideas about digital participation derived from them.

9.3.2. *Rushcliffe Borough Council's interim local plan consultation*

The local example involved Rushcliffe Borough Council, an English district authority, whose headquarters lie on the outskirts of Nottingham, in the middle of England. The Borough is largely rural and has a relatively affluent population, although pockets of poverty exist in former coal-mining areas. The research focussed on an interim local plan consultation, then part of English statutory plan-making.

The interim "review" consultation was designed to develop citizens' understanding of a housing allocation set by central government and Nottinghamshire County Council. It also aimed to gain the public's opinions before the more formal plan was established.

The plan's consultation process involved several methods. The authority used a leaflet delivered to every resident and newspaper advertisements to direct interested citizens to an e-mail address and website. The website included an online feedback form and a small amount of supplementary information relating to public meetings in the borough. More traditional (council-supplied) methods were also used, such as comment forms at public meetings and a feedback slip that could be detached from the leaflet. Many residents became engaged with the housing issue. Pressure groups formed because of the potential impacts of developments on house prices, as a form of Not In My Back Yard-ism (NIMBY-ism). Motivated citizens wrote letters rather than using the leaflet slip and the pressure groups sent "standard letters" to some residents, encouraging them to sign, add comments and send these to the authority. There were also two petitions organized by the pressure groups, altogether stimulating a relatively large number of responses (see Figure 9.2).

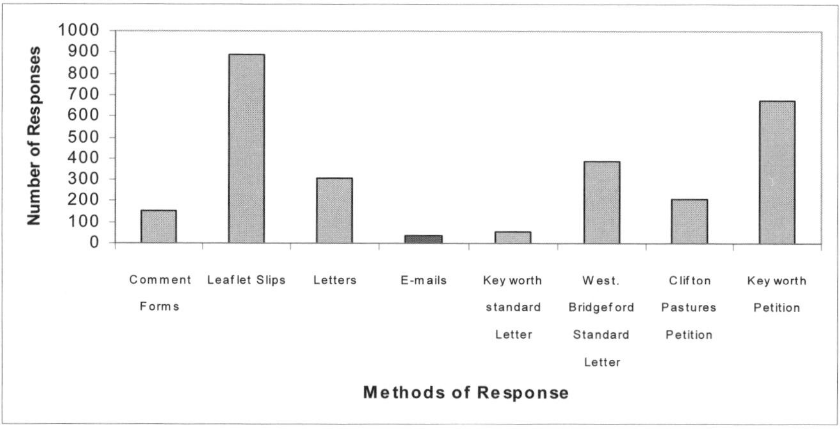

Figure 9.2. *Rushcliffe local plan responses [RUSH 99]*

This exercise generated around thirty formal responses through e-mail, a small percentage of the total number of contributions. However, it is also the example with the most online responses from a non-supported group of participants from the website survey. These low numbers offered benefits to both sets of actors. Participants could "discuss" ideas with the consultation's officer before making an official contribution, helping to ameliorate concerns and lower the response rate slightly. It also allowed the officers to manage their workloads more effectively than telephone enquiries.

The selection of different methods can be explained in a number of ways. Firstly, the standard letters and petitions are specific to certain geographical areas and reflect only a certain proportion of the population. This method would be more convenient to some participants rather than compose their own letter, increasing the numbers using this method with local support and encouragement. These letters helped citizens to identifying themselves with a particular NIMBY group, possibly adding weight to the issue through the numbers responding. In contrast, others may have valued the feedback slip on the leaflet as it was immediate and could be seen as the "official" response mechanism. Official weight can also be attributed to instances when participants chose to write letters and the different reasons given for using e-mail. Interviewees utilized the digital methods because it was seen as "cheap" and "immediate". They offered some people an opportunity to participate where it would have been difficult to before, either because of personal commitments (such as working times) or caring for others.

9.3.3. *The City of Edinburgh Council's first community plan consultation*

The strategic policy example comes from a unitary authority that delivers services to the Scottish capital, the City of Edinburgh Council. Like any European capital, Edinburgh has a historic core, provides the focus for much of the nation's economy and has a mixture of wealth and extreme poverty, although at times in quite close proximity. During the period of the research Edinburgh also became the location of the new Scottish Parliament, where a number of (electronic) democracy projects were being developed. The online consultation in this case was designed to generate additional responses from citizens and various representative groups. These voices would contribute to a pilot community plan that was being tested alongside four other Scottish local authorities. This strategic form of policy-making involved key local actors and the council creating a partnership that would actively involve citizens. The partnership's "visions" established how Edinburgh could be developed in a broad sense, drawing in areas such as sustainable development, decentralization of council services and urban regeneration. The main activity involved the plan partners writing documents for public scrutiny and consultation. The methods adopted included online versions of all the consultation documents, newspaper advertisements, a dedicated e-mail address, several meetings with business and voluntary groups, a call centre for people to telephone their responses (to cater for people who could not write English) and a telephone survey of their citizens panel. In the UK, citizens panels are democratically, demographically and statistically representative groups of around 1,000 citizens who are regularly consulted about council services and policy, often as part of Best Value service delivery policies. Edinburgh's approach did not as readily engage the general population and can be seen as more passive than the other cases. This was partly reflected in the response rate through the main consultation methods (see Figure 9.3).

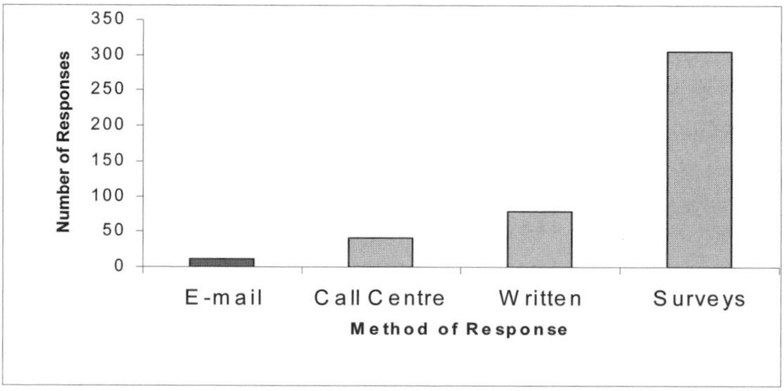

Figure 9.3. *Responses in Edinburgh's community plan [CITY 99]*

There were around twelve e-mail contributions from citizens in this example, with overall response rates much lower. Surveys were most effective as they targeted participants and asked them to respond immediately. Most other methods involved citizens framing their responses in structured text or at least structured speech that would probably have required some preparation.

Differences between the overall numbers responding in these two similar cases can be attributed to two qualities. Firstly, the issue in Rushcliffe was readily understood by residents and would have immediate and visible impacts on many citizens. In contrast, Edinburgh's strategic policy required additional interest and interpretation of the many issues at stake. This can be related to ideas about poorer responses to strategic policy-making in planning compared to local activities [THOM 96]. Secondly, Rushcliffe's publicity materials (the leaflet and press releases) were targeted and guided a particular group of participants, residents. In contrast, Edinburgh wanted the views from a very broad audience and their publicity materials may not have reached the intended target. As such, the relationship between traditional and digital methods is an important consideration if the numbers of people responding to an exercise is of interest.

9.3.4. *The London Borough of Lewisham's Dialogue Project*

The mixed policy example involved The London Borough of Lewisham, an English unitary authority comparable to Edinburgh's but with similarities to Rushcliffe as it exists in a larger government area, as noted above. The areas of Lewisham closer to the city centre have a great deal of poverty and large proportions of the population are from ethnic minorities. The authority felt that Lewisham residents were less well placed to have access to the Internet and this may have been the case compared to Rushcliffe or Edinburgh. However, Lewisham was also an authority that had an active interest in methods of participation and in evaluating their public participation activities.

Through their contribution to the Dialogue Project, the Lewisham case perhaps remains as one of the most technically advanced examples of digital participation. This Information Society Project Office (ISPO) initiative had partners in Ronneby (Sweden) and Bologna (Italy) and aimed to support and involve citizens who would not typically have access to the "information society". Lewisham's approach included training and the supply of free Internet access and some computers to around sixty citizen panel member's homes. The participants were involved in a series of online consultation activities, including consultant-facilitated chatroom debates about strategic issues with direct local impacts (hence the mixed position of this case). Topics included education, community safety and the future needs of Lewisham's elderly. Importantly, elected members then directly used these opinions to inform their decision-making. The council also encouraged the citizens to use e-mail (widely and between participants), bulletin boards (to leave messages about topics of interest) and to explore other online resources, either provided by the authority or through the World Wide Web. This last area was aimed at developing citizens' opinions that they would share and express during the chatroom debates. There was no measure of participant activity during the online discussions but the project's evaluation literature recorded how useful participants found certain methods (see Figure 9.4).

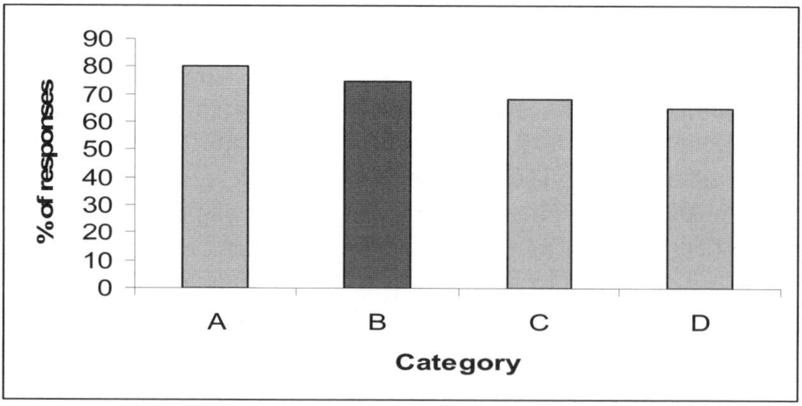

Figure 9.4. *Lewisham users' attitudes [LOND 99]*

Citizens had a slight preference for face-to-face contact (A) and e-mail (B) to the more involved (and technically difficult) tools such as chatrooms (C) and bulletin-boards (D). As well as providing a platform for debate, the technology also allowed participants to make new friends, learn new skills and contact family members in Australia and Canada, for example. The project demonstrated that those not normally familiar with digital methods could participate online and officers felt that contributions from the citizens did impact on councillor decision-making. As such, the Lewisham example was the most interactive, technically developed and well-supported case from this research, reflected in the authority using their own budgets to sustain the project for several months after ISPO funding had finished. From these cases and contexts stem ideas about what "public participation" is composed of in these top-down local government-driven contexts.

9.4. Components of public participation

The previous discussion illustrates that public participation has many forms constructed by actors throughout the participatory process. Five "components" can be identified that underpin these activities for both those inside and outside organizations or activities. The first of these components is an understanding of what "participation" means and the notions that relate to it, in a theoretical or philosophical way. These can often be related to ideas or ideals of democracy and democratic theory offers explanations and descriptions of them. They also help to construct and at times underpin the remaining four components. The second component reflects the "issues" that those who initiate exercises (for example officers) are willing to present for public scrutiny or concerns that citizens wish to see their local authority act upon. The latter is less common in most government-led participatory settings. Thirdly, many actors see the "audience" as an important feature. This not only involves the public who choose to participate (or not) and extra-organizational actors such as Central Government but also an internal audience of other officers in a local authority or politicians who may have expectations about the "outcomes" of an exercise. Outcomes themselves represent the fourth component. These involve perceived and actual results of exercises that occur at varying stages in a process. The perceived outcomes can influence what issues are made available for participation and those issues that citizens feel able to contribute to. Similarly, initial outcomes can also impact on directly related activities (including subsequent participation activities) or what actors have learnt from a participatory exercise both inside and outside of an authority. The final component that links others and provides an object to study is the methods of public participation. These range from traditional methods, such as a simple leaflet, to complicated digital ones, such as chatrooms or GIS. By examining the cases and these components it is possible to understand more about the complexities of participation itself through the lenses of these components before considering the role of GIS.

9.4.1. *Public participation in a local context*

In Rushcliffe, participation was a semi-formal exercise in gaining public support for an issue presented by the county council and central government. The authority wanted citizens to voice their concerns before the formal deposit consultation took place a year later. Officers wanted the public to have an interest in the tightly defined issue. The numbers of opinions expressed were important and the authority welcomed several methods of response. The exercise benefited from attempts to directly contact every resident through a leaflet that was delivered to every home.

The corporate view of participation came from an infrequently read officers' consultation guide. However, the tradition of consultation through the plan-making process is likely to have had a greater influence. Officers themselves viewed participation as a democratic ideal and a potential arena for confrontation. It was a place where issues could be dealt with but also one where caution was needed if the authority could not meet public desires. Officers felt it was important to find an appropriate audience and to present issues that citizens could understand or learn. This follows ideas of J.S. Mills that participation is educational [PATE 70] and that the activity in Rushcliffe would teach residents about their planning system.

Rushcliffe's residents were keen to see public participation as a means to have their (collective) voices heard. Some of the citizens interviewed, however, felt that since decisions had been made before arriving at Rushcliffe then the consultation was merely to foster good "public relations". Both groups of actors noted different types of "participant". For the officers it was someone who was "representative". For the citizens it included those who were too vocal in meetings, suppressing the ability of others to contribute. The authority recognized some barriers to access/participation that the typical plan consultation exercise generates (such as the technical and legalistic language involved) and attempted to overcome them. However, the "distance", in many senses, between actors meant that no real element of "trust" could be fostered and many officers and citizens did not feel that they had achieved all they set out to do.

The issue under consultation in Rushcliffe was very much a local one. The impacts of the consultation would ultimately shape the settlement and housing patterns of the borough and would impact on Nottingham's urban fringe and related settlements. The issue was fixed, the number of houses that had to be built was fixed and the majority of the consultation was very much about where these houses should be placed. Officers felt satisfied that residents had become motivated by the issue and responded in more numbers than during a previous similar exercise. In some ways the digital methods had provided two features. Firstly, they stimulated local interest by adding a fashionable feature to the exercise. Secondly they genuinely appear to have offered some residents an opportunity to contribute where a number of barriers preventing them from doing so had existed before.

This case had a specific target audience. The authority only wanted the views of residents. This involved the majority of citizens in the borough, a group that would directly feel the impacts of development either positively in terms of housing provision or negatively in terms of property values. The internal audience of the council involved the group who managed the website and dealt with authority-wide publicity matters (although the content came from the consultation officers). A number of officers from different departments had been actively involved in the consultation process, with their opinions and aspirations influencing the authority's overall approach. Rushcliffe wanted responses from as many people as possible. The leaflet was seen as a method to inform the population about the exercise and the proposed dwellings. The methods of response came from both the authority (leaflet-slips, comment forms from meetings and e-mail) and the community (the standard letters and petitions).

The authority's desired outcomes of the exercise were citizen-support for the local plan and an understanding of where residents felt new houses were needed. Participation would educate the public in planning and they would not feel distant from the decisions made by the authority. The actual outcomes were that a lot of people became motivated but there was an incorrect belief that agendas had already been set and that public influence was minor. The technology in the exercise was partly introduced as an after-thought and concerns were raised about how different contributions (e-mail, letter, feedback slip, petition etc.) would be handled, possibly influencing future activities.

The underlying notions of participation and desired outcomes in the organization shaped the consultation exercise. Officers wanted as many people to contribute as possible. They were concerned about "representative" participants and to have their concerns brought onboard and dealt with early on. This would lead to greater support for the full consultation a year later. The methods used were thus many and varied, with some being supplied by the community. In contrast, citizens gave a number of reasons for participating, varying from NIMBY-ism to being duty-bound to participate. The authority had an active and strong position in influencing the methods used, who the external audience was and what the outcomes should be.

9.4.2. *Public participation in a strategic context*

Edinburgh's exercise was structured around the democratic validation of an already set policy document. The community plan partners set the tone and content of the "vision" and looked to participants to indicate omissions, additions or errors. Given the breadth of the topic, participants indicated a number of issues. However, not all concerns were seen as suitable (as they may be a service rather than a policy issue) and would not be acted upon in the context of the plan. For this reason, the authority set some specific questions for the plan to be answered by participants. Also, although the numbers participating were important to the authority, attempts were made to involve the whole population and newspapers were seen as a means to encourage participation.

As those officers involved with the exercise came from the core of the authority they were very familiar with the council's corporate views of public participation. Several documents and support structures indicate the positive way Edinburgh wants to approach public participation. Logistical problems in the consultation exercise seem to have produced a form of participation that was not ideal in the eyes of the officers. In particular, their core value of "partnership" may have been active between the plan partners but less so between the partners and citizens. This notion was never mentioned by either set of actors interviewed in Edinburgh.

Officers also understood participation as an ideal and how this tended to frequently dissipate in practice. They noted that more ideal forms of participation were considered but that these may not have generated the wide-spread support for the community plan. Representation was important in this case. It was suggested that if there had been a large response to the consultation a sample of comments would have to take place to handle the range of opinions expressed. This may appear to have some statistical legitimacy but participants could have felt unhappy if they had participated and then entered a "lottery" to have their voice heard. Officers also were concerned about over-consulting some community groups that would lead to "consultation fatigue". Similarly, the citizens interviewed also saw participation in terms of ideals and practice. They also recognized the relationship participation can have with democracy but mistrusted organizations where opportunities to participate only occurred through elections, welcoming the chance to respond to the community plan.

The issue under consultation was a strategic one and the starting point for what was envisaged as a long-running policy process. The plan aimed to express a general approach for the way Edinburgh would develop over the next quarter of a century, and not only in terms of land-use. There was no problem or concern in a specific sense that the authority needed to address. In many ways this lack of specificity was the reason for few contributions from any of the audiences.

The external audiences for this case were very varied, reflecting some officers' notions of participation. Several representative groups were approached to obtain a range of opinions. The lay public's views were seen as valuable but additional. The ideas of being representative were reflected in actively looking for the opinions of four particular groups in meetings: the business sector, the voluntary sector, other public agencies (or other plan partners) and community organizations. Other potential participants were to contribute in less defined ways. Internal audiences were concentrated at the corporate core, although geographically speaking there offices were distributed across the city. Sometimes, there was an organizational distance too, with corporate responsibilities being held in devolved ward-based service departments. Officers' opinions were quite focussed through the plan but variations in their responsibilities led to different understandings of what participation was and how the other components came to bear on it, such as the importance of communication or a less than ideal form of participation relating to the methods adopted.

In Edinburgh there was also a mixture of traditional and digital methods. These were mainly used to gain opinions from groups through the consultant-mediated meetings and to deliver online information to participants. The citizens’ panel survey again reflected this representative approach and other methods used for support. Interviewees noted that these other methods performed less well and that there was a relationship between the response rate and the exercise’s low level of publicity. In particular, the prominence of the digital methods’ addresses (both for websites and e-mails) in the consultation materials was seen as inadequate, with the format of the e-mail address appearing to be unrelated to the authority and off-putting to potential participants. Although poorly performing, the e-mail was seen as more successful than the telephone call centre, which did not have the technical barriers of the Internet. Different groups used different methods on varying occasions. However, some participants contributed several times through different methods when they were performing different roles (such as by being both a transport issue lobbyist and a resident’s association member). This simultaneously muddled their representative nature and went against officers’ notions, noted above. In some senses the approach in Edinburgh had seen the technology as too much of an add-on and after-thought, where the implications of its adoption had not been fully considered, even though this was, notably, a leading example from the website survey which has highlighted the importance of looking behind website façades.

The scale of the issue and the methods used to focus on representative groups in this case can be contrasted with the first example. However, Edinburgh also wanted to encourage responses from general citizens and there are a number of similarities between these first two cases. The different issues, however, created different relationships between audiences, methods, the underlying notions of participation and what were eventually the outcomes of the exercise.

9.4.3. *Public participation in a mixed decision-making context*

The nature of engagement in Lewisham relied on close supervision of the participants and support for their activities. As part of the "Lewisham Listens" policy, the Dialogue Project was designed to aid councillor decision-making. As such, there was a limited possibility for citizens to participate in the upper "rungs" of Arnstein's ladder. Information was actively supplied to participants and they were encouraged to respond through a variety of media, including chatrooms. The consultees were already involved in council-based decision-making through the citizens' panel. Like Edinburgh, Lewisham had several documents that indicated the importance attached to meaningful public participation. Lewisham Listens was the focus for these activities but officers who ran the Dialogue Project went further than many authorities in trying to achieve participatory practices that promoted equality, transparency and openness, which were felt to help tackle social exclusion in deprived areas. The authority wanted the results broadly disseminated through a magazine delivered to every resident in the borough. Their interest was also characterized by their evaluation of the impact that public participation had on councillor decision-making. The officers' understanding of participation was typified by many citing Arnstein's ladder as "shorthand" for the different forms of participation that can occur in practice and that this could often be less than the ideals that the framework supports.

The citizens involved also supported participation and became heavily involved with the exercises. They, however, ultimately felt they were not being listened to. Perhaps they did not realize their role was to inform councillors but this should have been less of a concern as the elected members frequently chose to follow participants recommendations. Like the officers, many citizens viewed participation in terms of a theoretical ideal, related to democracy, but saw flaws in practice.

The nature of the issues under consultation was mixed because the participants were asked general opinions of issues that would have direct impacts on services in the borough. For example, the issue relating to the needs of Lewisham's elderly would have an influence on the provision of special accommodation for older citizens. Most participants could readily appreciate this issue but some of the younger participants could not readily engage with this topic and overall participation by this group was low across panel activities. The other issue in this case was to examine people's ability to use the Internet who would not normally have the opportunities to do so before.

As such, the audience was made up of individuals from a variety of demographic backgrounds. These panel members were specially selected and received a lot of attention, training and support throughout the project. It demonstrated that a variety of people could readily engage with the technology and participate quite actively. This project also meant that less activity was needed to engage citizens from the outset, unlike the other cases. In contrast, the internal audience of the exercise was made up of a very small group of officers at the corporate core plus an external consultant. Most of the interaction between participants and the authority occurred through the consultant, an administrative or technical officer. The Dialogue Project had a relatively high profile in the authority and strong support from senior officers and politicians.

The methods in this case were almost wholly digital, apart from occasional face-to-face meeting. Technical difficulties were noted such as poor Internet connections and a variation in participant abilities to type in a chatroom "conversation". However, for some participants, the online methods were very important as they offered them the opportunity to contribute where it would have otherwise been difficult to do so because of mobility problems or anxiety about crowded public meetings, for example. Technology was not an after-thought for this case but very much at the heart of the Dialogue Project.

The outcomes of the exercise are different from the other cases, partly as a result of its ephemeral nature. Although a great deal of support was given to citizens, they did not feel they had participated fully or were not being listened to, even though strong evidence exists to the contrary. This has also to be viewed in terms of the participants' sense of satisfaction and enjoyment of the project. The educational benefits may have tarnished some views of what public participation is but readily recognized ancillary benefits included making new friends across generations, contacting relatives online and learning transferable information technology skills. These outcomes are valuable and something that is often overlooked by the evaluation literature of many participatory exercises which may focus a great deal on power relations between actors. However, the authority has now ended the project and it is unclear what has happened to the participants, their skills and the impacts on ongoing e-government and participatory activity.

In the Lewisham example, participation was very much about experimenting with an informed citizenry to aid political decision-making. The methods and the audience were closely selected as part of an experiment to discover what the activity could provide. The issues under consultation, however, were not attractive to all participants and it is possible that the outcomes of the exercise could have been improved. For example, if citizens could have had a better understanding of their roles to inform politicians.

Examining these three cases has started to "unravel" what public participation is in digital and traditional settings and can demonstrate the interplay between its parts. Many of these components are interdependent. A strong relationship occurs between notions of participation, certain issues and the desired outcomes from any exercise. For some officers, the deployment of certain methods for a certain audience often seems to be less crucial than the numbers involved. This becomes complicated when participants choose to respond several times, in different guises for different "representative" groups. Additionally, actors were less concerned about the poorer performance of the digital methods, looking past them to factors beyond the control of the authority. Methods of participation are the point of interaction and notions of participation are embedded in them. Notions and outcomes appear to drive an exercise from the conception (and perception) of an issue, through articulation in the methods used for a certain audience, to the outcomes (desired, perceived or otherwise). Participation is readily identified as a theoretical entity and a practice by many of those involved.

The complex interactions between these key components need to be fully understood by local authorities engaged in these activities and those investigating them. There is also a need for appropriate policies and strategies so that officers' and citizens' expectations are clearly managed, nurturing a relationship of trust. Public participation is not a unique and shared construct, and achieving meaningful participation remains a significant challenge. Like digital methods, participation itself cannot be seen as an add-on to other activities but, in many instances, needs to permeate them all, potentially impacting on related activities involving geospatial data and analysis.

9.5. Public participation GIS and components of access

9.5.1. *GIS and access*

Having illustrated some of the complexity of public participation it is possible to focus attention on the role that GIS can have. Given the previous discussion of diffusion and the emerging policy context, it is perhaps unsurprising that this technology is being used as a method of participation, potentially offering something more developed and understood than just interaction through text. For example, the sense to which local actors more readily understand local rather than strategic situations (discussed above) presents advantages to utilising placed-based information. In addition, many issues contain a sense of location or "place" as part of the issue under consideration. Such ideas operate at differing scales, something which is primarily a geographical concept. Clear examples can be found in areas with a tradition for public participation such as land-use planning, with development control consultations focussing on a particular building;

regeneration projects tackling neighbourhoods and development plans covering entire local authority areas (as in the Rushcliffe case). This sense of scale may in turn impact on the types of participant who are likely to become engaged, as Thomas notes that the more strategic a policy, the fewer participants [THOM 96]. Therefore, the further away from the neighbour/neighbourhood setting the less likely the majority of citizens will respond (until an issue has national or international resonance). GIS also has a potential role in drawing together such place-based ideas and representing them in ways that can be readily shared and accessed in online environments. The ability to access information plays an important role as a notion of public participation, with artefacts such as maps offering special types of access and information worthy of particular consideration.

As much as our understandings of public participation can vary, so can ideas about access, especially for those involved in participation. In the UK, most participatory activity takes place through local authorities interested in citizens' views but equal interest in obtaining information can come from a public desire to be informed or to make an authority aware of an issue. The ways in which these two flows of information vary can be seen as "access" and "participation", depending on how far actors' roles are taken. Some citizens may have to seek information through complicated networks (both social and technical) and, therefore, encounter barriers to access. Others may receive guidance through this network or have information "pushed" to them through participatory activities. Local authorities may also face barriers such as obtaining a "representative" voice or who they want to consult. These demonstrate what forms of participation can take place, or what the intention is of those who are running participatory exercises. In as much as participation is varied, so are ideas about access to information and the two can be seen to have relatively identifiable inter-relationships.

This concept particularly emerges when examined from a geospatial point of view. Complicating factors such as heterogenous (social) conditions distributed over policy areas present difficulties and opportunities for public participation. For example, in a rural setting the dispersion of the population may make it difficult to draw together people to make their voice heard. Many look to the Internet to increase access to policies but this should be placed in an understanding of the urban-rural digital-divide, as recognized in European policy. The Internet is also seen as a way to offer "twenty-four-seven" (all day every day) access. However, the complexities of digital methods, including GIS, may present barriers to some accessing processes while opening avenues for others. For example, map-reading literacy is varied across cultures and all methods of participation include and exclude different groups [ALTY 87]. GIS is no different.

9.5.2. *Components of access*

From the above components of participation it is possible to explore their application to access to (geospatial) information. Those that initiate exercises (mainly local authorities) seek information from and provide information to potential participants, while participants provide responses or requests for information. These can vary considerably and influence the form of transaction that can take place, simultaneously encapsulating ideas of "access" or "notions of participation". A new framework would, thus, create "notions of access". When the purpose of access is related to participatory activity then notions of access become a special form of notions of participation. As such, the concepts behind audience and methods (of access) directly reflect those for participation but outcomes and issues need to be viewed slightly differently. Outcomes include the varying levels at which access is satisfied for all those involved: the ability to obtain the desired information. In some instances participation's "issues" could be maintained but access does not necessarily involve decision-making processes. For example data could be part of an information-sharing project between organizations or researchers. As such, "issues" would become "purpose", why access is sought and why those with the information would be willing to accept or decline a request. The framework must also, however, include what information is held and, therefore, a sixth component of "content" needs to be added to the framework. This is closely aligned with any methods employed, where the role of metadata becomes apparent as a means to indicate what information is available. In order to explore this issue further each component is again examined in turn.

Three different categories of access need to be examined regarding participatory activities that reflect technical, social and political aspects. Access has not been seen solely as access to digital infrastructures, although these are important; with the variation in access to telephone lines between rural and urban communities as a major part of that particular digital divide. It also involves an individual's ability to access "networks, machines, information and know-how of IT" [DUTT 99], which is taken further by Kling [KLIN 99] in terms of technological and social access. Technological access involves "the physical availability of suitable equipment, including computers of adequate speed and equipped with appropriate software for a given activity". Social access comprises of a competence to operate the equipment, reflecting "a mix of professional knowledge, economic resources, and technical skills to use technologies in ways that enhance professional practices and social life".

In a digital participation context, ideas of access have to be extended to include ideas of power within a democratic framework. Therefore, it is not only one's ability to access and understand the information but also to make effective use of it. This could possibly be through leveraging power and trying to affect outcomes. Smith and Craglia [SMIT 03] suggest that this could be referred to as "political access", which could also encompass the concept of an "equality of rights" – that current social and economic inequalities can undermine "formal equality" where the "legal right" to participate is understood [PERC 95]. This may be particularly true where the differences between the sides of the digital divide are rapidly increasing.

As with participation there is an internal and external "audience" for certain types of GI. From the "top-down" perspective, local authorities may not have a clear idea about who wants the information they hold. Those who wish to use local authority data may have very different backgrounds, abilities and demands for information at varying scales. This can be seen through the strategic context of Edinburgh's community plan and the Datashare Project (http://www.datashare.org.uk). This web-based GIS contained mapped policy areas from different community plan partners so that users (with different levels of access) could see which areas in the city had an over- or under application of resources. Importantly, this included citizens and formed an early example of a local SDI, of sorts.

From this top-down perspective it may be inappropriate for an organization to try to produce data in ways that meet the competing demands of all audiences, particularly when finances are heavily restricted. The Internet has offered increased access to information through websites (and portals, as seen more recently through INSPIRE's developments) and the ability to contribute to activities outside of normal office working hours in ways that often suit both the participant and the authority. However, the inequalities produced by the digital divide and varying public access offered by authorities to citizens impact on participation. As a result, metadata has emerged as a useful way for both external and internal audiences to know what information is available or has been obtained as a first step to increasing access, alongside the other social settings highlighted though the components of digital participation.

Unlike participation, access to GI focuses more on obtaining information rather than the methods used to get that information. The Internet and associated technologies have undoubtedly sparked actors' imaginations about increased information-sharing, access to data and the publication of local authority documents. In many ways ICTs are seen as the most suitable vehicle to deliver most of this information so it can be used on a variety of platforms, highlighting the usefulness of the digital participation research. When access is considered in relation to methods of participation then the role of mass communication becomes apparent. For example, citizens were contacted in Rushcliffe through several approaches to guide participants to the digital methods. In such contexts, the authority must improve universal access by providing materials that all citizens can obtain or respond through, relating to the notions of participation an authority wishes to follow. The call centre in Edinburgh was designed to meet this latter function as a more user-friendly means of response but it had limited use because the issue generated limited interest and it was seen as too informal. It is unclear how GI technologies could perform in similar circumstances, as most GI producers and consumers will prefer digital formats to cut down on the costs of replication or to draw on its integration abilities, for example. GIS remains a technical activity and often involves a specialist audience but experts could help citizens, as seen in advocacy planning, and the increase of public mediated online GI through, for example, Google maps is a noteworthy area of development. Any exploration of formal GI content is reflected in the role of metadata and catalogues that specify what information is present and what conditions are placed on obtaining it, either legally or financially. This can also involve ideas relating to political access, as there may be issues about knowing that certain information is available and how to use that information effectively.

Participatory settings often have identifiable outcomes for all those involved. The information provider, however, often does not see how their client has used their data, potentially impacting on some legal issues relating to liability and data licensing. There are several ways in which this subject can be tackled but it should be understood that it is the lack of access where many concerns should lie. This is particularly the case when an information monopoly exists that creates prohibitive prices or other barriers for certain (mainly public) actors. For example, some information may be affordable to a property developer but not for an individual citizen interested in tackling a planning issue. Barriers can also be related to some in organizations valuing information based on the purpose for which they have collected or analysed it but less so for its reuse. As such, certain information may not be made available for public consumption because its potential role is not recognized.

The final area is, therefore, the purpose for which people are attempting to gain access to information, something that an authority may require when judging if information should be released and at what price. People have many reasons for seeking information, such as research, product development, information-sharing between public bodies and public participation. Each of these has its own set of underpinning notions. PSI can be collected for a number of different reasons and a great deal of this must already be made readily available to the public by both English and Scots law in the UK, often stemming from European policy. How a judgement is made about this information's release is unclear and the variation in Freedom of Information laws between Scotland and the other parts of the UK may create an undesirable variation in access, particularly for SDI development. Purpose also, however, provides a focus for development and, more recently, infrastructures for geospatial information-sharing are emerging in thematic domains such as community safety and policing and inter-organizational data-sharing infrastructures, again harking back to the joined-up government ethos of e-government and the present initiative of transformational government.

9.6. Conclusion

Today, several features are coming together that will shape how we research and use certain technologies for public participation and access to (geospatial) PSI. The policy framework in the European context not only wishes to increase access and reuse of information held by public bodies but also identifies access as an important component of democratic activity at all levels of government. For nearly ten years, the Information Society policies of the European Union have been supporting the capacity of Member States to engage with this topic. GI has an important role to play, through underpinning the vast majority of this information, providing an economic case for increased access to it and a means to integrate policies from diverse sectors, from transport to agriculture. As such, national and regional SDIs have a role to play in providing a framework to integrate GI into e-government activities and allow people to know, both inside and outside organizations, what information is available. In addition, the environment, through the Aarhus Convention, European policy and the INSPIRE Directive, has been a key issue, where impacts move from local circumstances to major global concerns. Similarly, the UK has been developing the necessary physical infrastructure to support an information society since the early nineties. Policy developments have focussed on e-government, freedom of information, activities that support the continued improvement and democratisation of public services and attempts to make local authorities the core of civic leadership, all with an increased focus on the engagement of the public.

A challenge remains, however, in questioning the assumptions of this e-government activity, particularly as it applies to the increased involvement of citizens in local decision-making. One area of investigation has been attempts to find common themes of exploration between researchers in Europe and the USA for both access to GI and instances where such information becomes involved in participatory practices. A meeting in Spoleto, Italy, was very much a starting point, developing a core research agenda for access [WEHN 02] and participation [SMIT 02], and drawing together both researchers and their work through two special issues of the URISA journal (see: http://www.urisa.org/Journal/onlinejournal.htm). This book has continued this process, where the first edition exposed some of the issues to a francophone audience. The key themes from the access research agenda remain broad and have local-based information as a particular concern, where its institutional and organizational issues have clear applications to local authorities, their employees and their relationships with citizens. This can be seen to relate well to public participation and how those outside organizations may want to use GI. Access can be better understood in this context by exploring professional cultures and a government's role in SDI development, particularly through behavioural studies. Research topics need to examine how PSI can be (varyingly) accessed for democratic, research and economic purposes, as well as what legal and technical models are needed to expand meaningful access in diverse contexts. Comparative case studies and an investigation into business models have a role to play in developing appropriate approaches to the particular case of public sector geospatial information. Similarly, understanding varying geographies of access can explain actors' social contexts and their ability to access information, thus illustrating some issues of the "digital divide". A particular example of this divide can be seen in the inequality of access between urban and rural areas, potentially placing the unit of investigation, again, at a local authority level.

The specific themes of the Participatory Approaches Using GI (PAUGI) research agenda also involve local organizations and equally draw on a need to understand "access". Work over many years must be drawn together and presented in a variety of formats, from wider dissemination of findings in the academic community to publications that allow researchers and practitioners to share ideas in less formal settings. There is also a need to develop methodological frameworks that will help evaluate PAUGI projects, requiring a strong, trusting relationship between researchers and local actors. This concept of trust will also emerge when exploring the role that technology plays in local government settings, investigations into GI learning-environments and instances where local examples gain greater meaning at aggregated levels.

In terms of the discussion of this chapter, it has also been recognized that the meanings of "participation" in PAUGI contexts and the relationship between access to GI and public participation requires further research. The research discussed in this chapter indicates that participation is a construct that is neither unique nor equally shared. As such, five components outline what "participation" can mean to those involved. In particular, the notions of how participation can be viewed; those involved in participatory activities both within and outside organizations; the (traditional and digital) methods that can be employed to conduct participation; the issues that participation is suitable for; and the perceived and actual outcomes of participatory exercises all meet to describe and shape actors views of public participation and how it can be examined in practice (or shortly after). It has been argued that participation and access to information can be seen as existing on a spectrum of information sharing (where access is mainly unidirectional and participation multi-dimensional) and that these components can be adapted to begin to examine "access" to GI, with this concept also existing as a notion of participation in some instances.

As a result, actors' notions of access will be important in shaping what they expect to happen when a transaction for GI is attempted, both in terms of those with the information and those attempting to gain access. The notions are particularly important in participatory contexts as the ability to gain access goes beyond the technical and social forms identified by, amongst others, Kling [KLIN 99] but also an idea about using this information in ways that accurately communicate the message of participants or cause desired effects such as political leverage for funding. The "audience", as in participatory contexts, will take the form of an internal and external set of actors. Any barriers may be related to the translation of the "issues" component to one of "purpose" and the perceived or actual outcomes of allowing/obtaining access. The major change between these two frameworks is the need to recognize that "information" (in a broad sense) should be the main unit of observation rather than the methods employed. Through the additional component of "content", digital methods offer the principle avenue to secure access to most GI. As such, it is metadata that links these methods to the information content. Finally, these two non-exclusive frameworks require further empirical testing and refinement to expand our understandings of the relationships between access to information and public participation, alongside their particular application to GI, as both a facet of electronic (or transformational) government and a base for much of our expanding information society.

9.7. Bibliography

[ALTY 87] R. ALTY., R. DARKE, "A City Centre for People: involving the community in planning for Sheffield's central area", *Planning Practice and Research*, 3 (1), pp. 7-12, 1987.

[ARNS 69] S. R. ARNSTEIN, "A Ladder of Citizen Participation", *Journal of American Institute of Planners*, 35 (4), pp. 216-224, 1969.

[BRYD 98] C. BRYAN, Z. TSAGAROUSIANOU, D. TAMBINI, "Electronic Democracy and the Civic Networking Movement in Context", in Z. TSAGAOUSIANOU, D. TAMBINI, C. BRYAN (eds.), *Cyberdemocracy: technology, cities and civic networks*, London, Routledge, pp. 1-17, 1998.

[CITY 99] CITY OF EDINBURGH COUNCIL, *City-wide Community Planning*. Progress Report, June 1999, Policy and Resources Committee, Edinburgh, City of Edinburgh Council, 1999.

[CLAR 97] D. CLARK, *Your Right to Know: the Government's proposals for a freedom of information act*; http://www.national-publishing.co.uk/ document/caboff/foi/; 14/02/1998 1520h, 1997.

[CRAG 02] M. CRAGLIA, A. ANNONI, R.S. SMITH, P. SMITS (eds.), *Spatial Data Infrastructures: country rep rts 2002,*. Geographical Information Network in Europe; EC-Joint Research Centre (EUR 20428 EN): http://wwwlmu.jrc.it/ginie/doc/SDI_final_en.pdf.

[CRAG 03] M. CRAGLIA, R.S. SMITH, S. URBANAS (eds.), *Geographical Information Policies in Europe: technical report*, 2003 Geographical Information Network in Europe: http://wwwlmu.jrc.it/ginie/doc/ DP_TechRpt_Publish.pdf.

[DGXI 96] DGXIII/E, *Info 2000: Stimulating the Development and Use of Multimedia Information Content*. Luxembourg, European Commission, 1996: DGXIII/E.

[DUTT 99] W.H. DUTTON, *Society on the Line: Information Politics in the Digital Age*, Oxford, Oxford University Press, 1999.

[EURO 93] EUROPEAN COMMISSION, *White Paper on Growth, Competitiveness, and Employment: the Challenges and Ways Forward into the 21st Century* (COM (93) 700 final), Brussels, Commission of the European Communities, 1993.

[EURO 1a] EUROPEAN COMMISSION, *Citizens and Consumers,* 2001a: http://europa.eu.int/information_society/topics/citizens/index_en.htm.

[EURO 1b] EUROPEAN COMMISSION, *Education and Culture,* 2001b: http://europa.eu.int/information_society/topics/education/index_en.htm.

[EURO 1c] EUROPEAN COMMISSION, *eEurope Framework*, 2001c: http://europa.eu.int/information_society/eeurope/news_library/pdf_files/initiative_en.pdf.

[EURO 1d] EUROPEAN COMMISSION, *Employment and Social Affairs*, 2001d: http://europa.eu.int/information_society/topics/employment/index_en.htm 04/07/2001 1638h.

[EURO 1e] EUROPEAN COMMISSION, *Ten-Telecom: trans-European telecommunications network*, 2001e: http://europa.eu.int/information_society/programmes/tentelecom/index_en.htm; 04/07/2001 1628h.

[HUNT 99] D. HUNTER, (ed.), *The Connected Council: leading the community in the information age*. Bristol, Foundation for Information Technology in Local Government (FITLOG), 1999.

[KLIN 99] R. KLING, "Can the 'Next Generation Internet' Effectively Support 'Ordinary Citizens'", *The Information Society*, 15 (1), pp. 57-63, 1999.

[LOND 99] LONDON BOROUGH of LEWISHAM, *The Dialogue Project: consultation on-line*. London, London Borough of Lewisham, 1999.

[MASS 96] I. MASSER, H.J. CAMPBELL, "Great Britain: the dynamics of GIS diffusion", in I. MASSER, H. CAMPBELL, and M. CRAGLIA, *GIS Diffusion: the adoption and use of Geographical Information Systems in Local Government in Europe* (GISDATA Series No. 3). London, Taylor and Francis, 1996.

[OFFI 02] OFFICE OF THE DEPUTY PRIME MINISTER, *The National Strategy for local e-government*. London, HMSO, 2002: www.localegov.gov.uk.

[PATE 70] C. PATEMAN, *Participation and Democratic Theory*. Cambridge, Cambridge University Press, 1970.

[PERC 95] J. PERCY-SMITH, *Digital Democracy: Information and Communication Technologies in Local Politics*. Research Report No. 14, London, Commission for Local Democracy, 1995.

[RUSH 99] RUSHCLIFFE BOROUGH COUNCIL, *Local Plan Working Group: Borough Local Plan Review*, (22/07/1999) West Bridgford, Rushcliffe Borough Council.

[SCOT 1a] SCOTTISH EXECUTIVE, *Freedom of Information: opening up government,* 2001a: http://www.scotland.gov.uk/news/2001/03/se0501.asp.

[SCOT 1b] SCOTTISH EXECUTIVE, *Scotland To Have Its Own Freedom of Information Bill- Wallace,* 2001b: http://www.scotland.gov.uk/news/releas99_5/pr1076.htm.

[SMIT 01] R.S. SMITH, *Public Participation in the Digital Age: a focus on British local government.* PhD Dissertation. Sheffield, University of Sheffield, 2001.

[SMIT 02] R.S. SMITH, *Participatory Approaches Using Geographic Information (PAUGI): towards a trans-Atlantic research agenda.* 5th AGILE Conference on Geographic Information Science, Palma (Balearic Islands, Spain) April 25th-26th, 2002. See http://agile.isegi.unl.pt/Conference/Mallorca2002/Papers/pdf/dia25/Session_3/s3_Smith.pdf.

[SMIT 03] R.S. SMITH, M. CRAGLIA, "Digital Participation and Access to Geographic Information: a case study of UK local government", *URISA* (15) *Special Edition- Public Participation GIS* (Volume II Specific Participatory Experiences and Shorter Response Articles), 2003.

[THOM 96] H. THOMAS, "Public Participation in Planning", in M. TEWDWR-JONES (ed.), *British Planning Policy in Transition: planning in the 1990s.* London, UCL Press, pp. 168-188, 1996.

[WEHN 02] U. WEHN DE MONTALVO, *Access to Geographic Information: towards a trans-Atlantic research agenda.* 5th AGILE Conference on Geographic Information Science, Palma (Balearic Islands, Spain) April 25th-26th, 2002. See http://agile.isegi.unl.pt/Conference/Mallorca2002/Papers/pdf/dia25/Session_3/s3_Montalvo.pdf.

Chapter 10

GIS and Citizen Participation: Urban Revitalization Projects in the USA

10.1. Introduction

The last 10 years have seen an incredible increase in the adaptation of GIS in different types of organizations. They have also seen the evolution of participatory GIS (or public participation GIS), which is the subject of new research programs, involving access to geographical information technologies (GITs) by traditionally marginalized communities often forgotten in the computer revolution. In addition to access, research in terms of participatory GIS also favors the exploration of the complex nature of the adaptation process of GIS, as well as their social and political impact. In practice, participatory GIS evolved worldwide through a multitude of applications and user communities. The most significant applications of participatory GIS can be observed:

– in the USA, in urban revitalizations and land-management programs partly taken over by the residents [BOS 02, CAS 02, ELW 02A, ELW 02B, GHO 02, GHO 03, SAW 02, VEN 02];

Chapter written by Rina GHOSE.

– globally, through management and natural resource preservation efforts [JOR 02, KYE 02, MAC 02, MER 02];

– as well as in different forms of activism organized locally, nationally and globally [SIE 02, STO 02, TUL 02].

The main goal of this chapter is to analyze the interlinked complexities characterizing participatory GIS projects in contexts of neighborhood revitalization of North American cities.

The participatory approach in the American urban landscape was mainly developed to help citizens and land developers in their revitalization efforts for underprivileged neighborhood communities. These areas have experienced the effects of a large and quick de-industrialization, followed by significant disinvestment, creating problems of poverty, unemployment, criminal activity, dilapidated housing and a quality of life that has deteriorated. Community organizations, representing the citizens of these underprivileged districts, are actively involved in these revitalization efforts, as well as in communicating the problems encountered daily to the proper government agencies. These organizations want increasingly a real say in the management of their neighborhood and want to collaborate actively with local community departments in all revitalization actions.

In their efforts and energy to participate efficiently and formally in the management activities of their neighborhood, these community organizations are faced with an increasing need for geospatial technologies and data. Unfortunately, access to this data, as with associated technologies, remains difficult for a variety of reasons. These problems limit the efficient participation of the citizens [BAR 94, SAW 96, CRA 98, ELW 98, ELW 02A]. On the one hand, these citizen groups often only have "very little financial resources" [SAW 96], limiting their capacity to acquire the necessary informational resources (data), software and hardware. The lack of means also limits access to appropriate education for their personnel who have no (or very limited) skills in the field (use of technologies or basic knowledge of spatial analysis, for example). On the other hand, these organizations have to deal with an additional obstacle. The high turnover generates higher training costs [BAR 94, BAR 98, BAR 02]. Finally, access to public databases is another challenge for these local citizen groups. And yet, these data are often an important knowledge base of the different aspects of the operation of their community and their areas of life.

These problems are at the basis of a technological division between official land-management agencies (administrations and local departments), who are savvy GIT users, and local citizen groups for whom GIT is particularly complex to use. Social disparities are increased in this way by the inequities in accessing "geographic" knowledge and associated information technologies. Because, in fact, accessing such knowledge and know-how is political, economical and social power for local administrations, compared with the citizens of these disaster areas who are often helpless [CAS 96]. The assessment of this new form of "digital" relegation generates concerns and questioning from some professionals, intellectuals and academicians involved in the reflections of new social public politics [HUT 93, BAR 98, HAR 98, OBE 98, LEI 00]. In the last decade, teams of researchers were grouped with government and non-government agencies to implement participatory GIS initiatives to support access of GIS and spatial data to marginalized groups. Since then, an unprecedented number of community organizations has adopted and used this technology for land management, problem resolution and distribution of services [CRA 98, KEL 99; GHO 01A, GHO 01B, GHO 02, SAW 02, GHO 03]. However, case studies carried out in the field show that these experiences are anything but simple and linear. In fact, they are complex and multi-dimensional processes that should be explored. Another reason they should be explored in this book is that these participatory GIS experiences not only constitute very relevant situations of analysis of "GIS and society" relations, but they also enable us to address organizational dimensions (citizen organizations) specific to geomatics.

10.2. The role of local contingencies in participatory GIS

Participatory GIS projects greatly depend on the context, as their efficiency and sustainable development depends on a certain number of local random factors. Case studies that I carried out in three large American Midwest cities (Minneapolis, Milwaukee and Chicago) show how local contingent factors are vital to the implementation, sustainable development and efficiency of participatory GIS initiatives in community organizations [GHO 01A].

The citizen participation process in the urban-management policy led by local government is a first key element to determine the efficiency of participatory GIS. Even in the case, where this type of government is open to cooperation, political processes involved remain extremely complex. Participatory GIS then greatly depends on the good will of local communities to involve community organizations and to grant them a certain authority in the context of formal urban land-management practices. Consequently, the cities that formalized strategic neighborhood land-management processes involving residents (Minneapolis or Milwaukee) could observe a significant increase in the participation of the citizens, as well as a more important deployment of energy from community organizations toward participatory GIS. Conversely, a city in Chicago, only offering rare occasions for citizens to collaborate, and with no formal method of participation, has significantly less activity concerning participatory GIS from its community organizations. We must understand that formal land-management processes also impose some constraints: for example, a series of conditions and instructions to which community organizations must comply. It can be difficult for these organizations to apply such conditions in view of their employee turnover and lack of funds. Community organizations can also be lacking in knowledge or expertise to efficiently develop an urban land-management project. In addition, such projects often carry priorities imposed to community organizations, who then have to put aside their own objectives to satisfy the priorities chosen by institutions.

GIS experience and expertise from communities and local administrations, especially in terms of urban land-management practices, is a second factor to consider when encouraging participatory GIS projects. The existence of this type of experience often generates a clearer awareness of the value of using GIS in land-management programs. In this way, research reveals that local communities with this type of experience support community organizations involved in participatory GIS initiatives better: for example, the local administration responsible for the financial and informational aspects to share commitments and actively encourage these initiatives. Milwaukee is a perfect illustration of this phenomenon. In fact, municipal services and local administrations are pioneers in the USA in terms of urban GIS development and use. Aware of the stakes, they quickly and constantly supported the initiatives of participatory GIS launched by community organizations. The local branch (Community Block Grant Organization) from the federal "Housing and Urban Development (HUD)" department directly encouraged the development of participatory GIS by associating them to their own strategic urban land-management problem (Neighborhood Strategic Planning). This encouragement is all the more important, as this program is part of a federal grant for community organizations to help them get involved in central neighborhood revitalization. This federal service also financed the creation of a non-profit association dedicated to providing free and personalized GIS analyses to community organizations. In concrete terms, this support and encouragement is the direct cause of massive expansion of GIS use by citizens' associations in Milwaukee.

The existence of a solid network of stakeholders or organizations (associations), qualified as intermediate, is another crucial factor for the development of GIS initiatives in community associations with few resources. In fact, they must often be creative in trying to commit to several cooperations that a GIS initiative requires. These intermediate stakeholders ensure contact with larger organizations: private foundations, national non-profit organizations, local universities, local and federal administrations, etc. These organizations generally provide spatial data, GIS technologies, computer equipment and other equipment necessary to train personnel of organizations at low cost or even free. These organizations also commit in long-term partnerships to help community organizations in maintaining sustainable use of GIS. In practice, these intermediate stakeholders are at the basis of successful participatory GIS initiatives in Milwaukee and Minneapolis. In the case of Chicago, the importance of these stakeholders is huge, because communities and local administrations have very little direct and formal interactions with community organizations.

The willingness of communities and local administrations to share their resources (digital city maps, data on housing conditions or real estate appraisals) with community organizations or intermediate associations is another vital pre-condition to the development of participatory GIS initiatives. However, sharing of data often relies on a political process, and its logic is not always seen for the layperson. Case studies show how cities that have facilitated access to data have also encouraged citizen participation to land-management projects through participatory GIS. In more precise terms, local planning centers with the most contribution to public database sharing and supported spatial analyses with the help of user friendly Internet GIS, for example, offers greater opportunities of using participatory GIS. Throughout all these development, efforts for Web-based GIS; distance, time, cost and expertise barriers are not as important. These online solutions make frequent use of spatial data possible for community organizations in their daily activities. The Milwaukee and Minneapolis cases clearly demonstrate the importance of this factor.

10.3. Production and use of spatial knowledge in participatory GIS

Community organizations in the USA are actively working toward the acquisition, maintenance and use of spatial data. Our previous case studies highlighted the nature of the most widely used spatial data by community organizations. They mainly consisted of data on daily urban problems from neighborhoods: housing conditions, economic development, criminal activity and security, safety and environmental conditions. In addition, these organizations collect and update data concerning their members and potential clients (addresses, phone numbers, demographic characteristics, involvement in tasks and services from the community organization and services rendered). The acquisition of this type of data can be complex or relatively easy depending on the will of local administrations and communities and the network of intermediate stakeholders to share this data. In addition, community organizations are increasingly aware of the value of their empirical local knowledge. They are often involved in the collection and update of data from their respective districts. This neighborhood data offer them a lever for the local administration to initiate actions and participate in the remodeling of urban policies in some cases. This explains why community organizations are increasingly interested in using spatial data at different geographical levels: address, lot, building, district, region or the census perimeter. This multi-scalar approach provides them with a chance to be in a wider political and geographical context, and thus lead comparative studies for the possibility of examining the quality of their neighborhoods, in perspective with the whole city.

Research studies on public participation GIS (PPGIS) clearly highlight how community organizations use geographical information to produce spatial knowledge. They then use this knowledge in different ways in a strategic perspective: confirm and legitimize existing empirical knowledge to initiate actions and formulate strategies; monitor the conditions of urban neighborhoods and anticipate changes; plan organizational tasks and financing and recruitment efforts; produce new information to improve the service offering; explore spatial relations to question or remodel urban policy [GHO 03]. The major types of spatial knowledge production were identified in Minneapolis [ELW 03]. The end of this section reviews them.

Although community organizations have a wealthy heritage of local data, often based on their empirical knowledge, they still use public databases when they have to use a GIS to confirm and legitimize their knowledge. The analyses and processes from these databases provide new data to community organizations. Equipped with new proofs, they often strive to condemn the inaction of municipal services concerning their neighborhood problems. Thematic maps produced by GIS can also be a testimony of non-expert knowledge involving existing problems in neighborhoods, enabling community organizations to locate these problems and identifying models and trends. The process of verification of local knowledge compared to public databases can be the opportunity for community organizations to analyze the quality and accuracy of public data and to evaluate the relative quality of data coming from local knowledge.

Community organizations are also highly interested in video surveillance for existing problems in underprivileged districts, such as urban deterioration, criminal incidence or violations of security or safety rules. Experience shows that using spatial knowledge instead of GIS can be extremely useful for the implementation of this type of surveillance system. For example, the cartographical representation (particularly thematic) of spatial and temporal knowledge enables the members of community organizations to evaluate whether the strategies chosen for revitalization were useful or not. This type of spatial knowledge can also help in anticipating potential problems in the future, which could be treated in advance by developing strategic plans in neighborhoods. In addition, thematic maps (formalizing models or trends of existing problems in neighborhoods) are efficient tools of communication during meetings with neighborhood residents. In fact, the visualization quality of the maps produced by GIS makes it easier to understand the problems, at least, better than the tabular data presentations.

Community organizations also use their spatial knowledge to search for financing from different sources. In the USA, since the decrease of funding from the states and federal, community organizations must turn to partnerships with the private sector. Private operation and financing modes systematically impose that information be up to date, records with past results, obvious proofs of current needs, detail in the methods used to respond to those needs and expected results. Using GIS and spatial data is a competitive advantage for the production of these files. It constitutes a guarantee of success for obtaining financing, undoubtedly because the use of these technologies gives a solid professional character to the representation of community needs.

Within community organizations, a new trend consisting of using geographical information with creativity through the Internet is emerging, to build spatial knowledge that will improve the service offering and the organization of community activities. Some community organizations insert digital pictures and plans on their web sites to attract potential investors to their neighborhood. A community organization created a comment form on the Internet offering residents the possibility of reporting and locating each of the problems encountered in detail. The organization then takes digital pictures of these problems and sends all these documents by e-mail to the appropriate department of the community or the local administration involved. This strategy made it possible to initiate concrete actions and to ensure that their organization was recognized as an organized and efficient entity. Finally, the organization hosts an online discussion forum for its residents to extend the possibilities of exchange and communication. Because of this, the residents, companies and state civil servants circulate information, organize meetings, dialog online and send e-mail messages to the members of the community. The members of this community organization estimate that this form of cybernetic work facilitates access to data on neighborhood conditions and problems. In general, the organization categorizes data by address in its database, and then transposes it in a cartographical form to bring out the trends or models that can then be integrated in their strategic plan.

Finally, community organizations use increasingly complex GIS analyses in the production of spatial knowledge, which are then used to remodel urban policy concerning revitalization of neighborhood districts. Studies on complex urban indicators produced by GIS were carried out over long periods to highlight the low quality of life of residents in some underprivileged areas. This type of study was very efficiently used to remodel urban policies and creating new investment opportunities in dilapidated districts.

10.4. Local variation of participatory GIS development

There are considerable variables in the capability of local community organizations to support participatory GIS initiatives and to apply efficiently and creatively their spatial knowledge to urban revitalization projects. The community organizations with the most success in understanding the power of spatial data and GIS use are those developing and updating the greatest range of geographical knowledge, including formal public data and local empirical data. They generalize the use of this type of data in most of their organizational activities. These organizations become active stakeholders in the production of local knowledge and are therefore less dependent on institutional or private providers. There are community organizations, however, who participate just as much and are involved similarly in participatory GIS initiatives, but who cannot take advantage of these opportunities. These are organizations where the use of spatial data and GIS remains limited and marginal. They do not make continuous cooperation efforts with the intermediate stakeholders in the field of participatory GIS. In fact, they remain highly dependent on external GIS expertise providers, accessing analyses in client/server mode. This variation in the development of participatory GIS is explained by a certain number of organizational factors, characteristic of the way in which a community organization appropriates and gets involved in a participatory GIS project, as well as how it responds to specific local contingencies [ELW 03]. The following factors are particularly predominant:

– The organizational knowledge and experience play a predominant role in the capacity of a community organization to structure the development of participatory GIS. The following in particular are vital: the knowledge of technology (including its power and limitations); the knowledge of local opportunities that can support participatory GIS projects; knowledge of efficient local development strategies; and finally, knowledge necessary to obtain financial and technical resources.

– The organizational capacity to create cooperation relations with key stakeholders in other community organizations, as well as in public or private institutions is another key factor.

– The organizational stability is another important variable in that it influences the sectors of organizational knowledge and the formation of networks, defining the level in which these resources are retained and preserved within the organization. The organizational stability includes the duration of management, personnel turnover, constancy in organizational missions and financial support.

– The type and status of the organization can also be significant factors. In the USA, there are two major types of community organizations: the "Community Development Corporation" (CDC) or the "Community Based Organization" (CBO). In the CDC model, the organization gives priority to economic programs and housing development. It is involved in large projects of capital investments and greatly depends on local corporate and private financing. In the CBO model, the organization focuses more directly on activities of community organization and depends more on public funding. In the current economical climate, a CDC-type structure is economically and politically more advantageous because it has more important self-financing power. Local communities and private foundations tend to favor CDCs over CBOs today, insofar as they perceive the first model as being more competent and productive and consequently, the one offering a better chance for efficient participation.

10.5. Conclusion and directions for future research

Participatory GIS projects seem all the more efficient as the capacity of a community organization is favored to actively participate in the management and urban revitalization of its district. However, the problems must not be overshadowed. The development of a participatory GIS is a real challenge, in particular, for local citizen associations with very little resources. These associations are doomed to a constant struggle to reach the objectives linked to the organization of their community, to the offering of services and to planning tasks for their neighborhood. The efforts must be constant to be able to support participatory GIS initiatives among the more modest and less well-organized local citizen associations. In addition, continuous examination of important organizational and political factors for participatory GIS remains open. The production of knowledge in the field is a complex process, having to deal with local contingencies. Long-term research, based on comparative case studies, is necessary to improve the intelligence of these processes, as well as to develop a strong and complete theoretical structure.

10.6. Bibliography

[BAR 94] M. BARNDT, W. CRAIG, "Data providers empower community GIS efforts", *SIG World*, vol. 7, 49-51, 1994.

[BAR 98] M. BARNDT, "Public participation GIS: Barriers to implementation", *Cartography and Geographic Information Systems,* vol. 25, 105-112, 1998.

[BAR 02] M. BARNDT, "A model for evaluating public participation GIS", in W. CRAIG, T. HARRIS, D. WEINER (eds.), *Community Participation and Geographic Information Systems*, London, Taylor & Francis, 2002, 346-356.

[BOS 02] M. BOSWORTH, J. DONOVAN, P. COUEY, "Portland Metro's dream for public involvement", in W. CRAIG, T. HARRIS, D. WEINER (eds.), *Community Participation and Geographic Information Systems*, London, Taylor & Francis, 2002, 125-136.

[CAS 02] L. CASEY, T. PEDERSON, "Mapping Philadelphia's neighborhoods", in W. CRAIG, T. HARRIS, D. WEINER (eds.), *Community Participation and Geographic Information Systems,* London, Taylor & Francis, 2002, 65-76.

[CAS 96] M. CASTELLS, *The Rise of the Network Society*, Cambridge, Blackwell, 1996.

[CRA 98] W.J. CRAIG, S. ELWOOD, "How and why community groups use maps and geographic information", *Cartography and Geographic Information Systems*, vol. 25, 95-104, 1998.

[ELW 98] S. ELWOOD, H. LEITNER, "SIG and community-based planning: Exploring the diversity of neighborhood perspectives and needs", *Cartography and Geographic Information Systems*, vol. 25, 77-88, 1998.

[ELW 02A] S. ELWOOD, "GIS and collaborative urban governance: Understanding their implications for community action and power", *Urban Geography*, vol. 22, 737-759, 2002.

[ELW 02B] S. ELWOOD, "The impacts of GIS use for neighborhood revitalization", in W. CRAIG, T. HARRIS, D. WEINER (eds.), *Community Participation and Geographic Information Systems*, London, Taylor & Francis, 2002, 77-88.

[ELW 03] S. ELWOOD, H. LEITNER, "Community-based planning and GIS: Aligning neighborhood organizations with state priorities", *Journal of Urban Affairs,* vol. 25, 139-157, 2003.

[ELW 04] S. ELWOOD, R. GHOSE, "Public participation GIS in community development planning: Framing the organizational context", *Cartographica*, vol. 38, no. 3&4, 2004.

[GHO 01A] R. GHOSE, "Use of information technology for community empowerment: Transforming geographic information system into community information systems", *Transactions in GIS*, vol. 5, 141-163, 2001.

[GHO 01B] R. GHOSE, W. HUXHOLD, "The role of local contextual factors in building public participation GIS: The Milwaukee experience", *Cartography and Geographic Information Systems*, vol. 28, 195-208, 2001.

[GHO 02] R. GHOSE, W. HUXHOLD, "Role of multi-scalar GIS-based indicators studies in formulating neighborhood planning policy", *URISA Journal,* vol. 14, 5-17, 2002.

[GHO 03a] R. GHOSE, "Investigating community participation, spatial knowledge production and GIS use in inner city revitalization", *Journal of Urban Technology*, vol. 10, 39-60, 2003.

[GHO 03b] R. GHOSE, S. ELWOOD, "Public participation GIS and local political context: Propositions and research directions", *URISA Journal*, vol. 15, APA II, 17-24, 2003.

[HAR 98] T. HARRIS, D. WEINER, "Empowerment, marginalization, and community-oriented GIS", *Cartography and Geographic Information Systems,* vol. 25, 67-76, 1998.

[HUT 93] C.F. HUTCHINSON, J. TOLENDANO, "Guidelines for demonstrating geographical information systems based on participatory development", *International Journal of Geographical Information Systems*, vol. 7, 453-461, 1993.

[JOR 02] G. JORDAN, "GIS for community forestry user groups in Nepal: Putting people before technology", in W. CRAIG, T. HARRIS, D. WEINER (eds.), *Community Participation and Geographic Information Systems*, London, Taylor & Francis, 2002, 232-245.

[KEL 99] W. KELLOG, "From the field: Observations on using GIS to develop a neighborhood environmental information system for community-based organizations", *URISA Journal*, vol. 11, 15-32, 1999.

[KYE 02] P. KYEN, "Promoting local community participation in forest management through a public participation GIS application in Southern Ghana", in W. CRAIG, T. HARRIS, D. WEINER (eds.), *Community Participation and Geographic Information Systems*, London, Taylor & Francis, 2002, 218-230.

[LEI 00] H. LEITNER, S. ELWOOD, E. SHEPPARD, S. MACMASTER, R. MACMASTER, "Modes of GIS provision and their appropriateness for neighborhood organizations: Examples from Minneapolis and St. Paul, Minnesota", *URISA Journal,* vol. 12, 43-56, 2000.

[MAC 02] P. MACNAB, "There must be a catch: participatory GIS in a Newfoundland fishing community", in W. CRAIG, T. HARRIS, D. WEINER (eds.), *Community Participation and Geographic Information Systems*, London, Taylor & Francis, 2002, 173-191.

[MER 02] T. MEREDITH, G. YETMAN, G. FRIAS, "Mexican and Canadian case studies of community-based spatial information management for biodiversity conservation", in W. CRAIG, T. HARRIS, D. WEINER (eds.), *Community Participation and Geographic Information Systems*, London, Taylor & Francis, 2002, 205-217.

[OBE 98] N.J. OBERMEYER, "The evolution of public participation GIS", *Cartography and Geographic Information Systems*, vol. 25, 65-66, 1998.

[SAW 96] D. SAWICKI, W. CRAIG, "Democratization of data: Bridging the gap for community groups", *Journal of the American Planning Association*, vol. 62, 512-523, 1996.

[SAW 02] D. SAWICKI, D. PETERMAN, "Surveying the extent of public participation GIS practice in the United States", in W. CRAIG, T. HARRIS, D. WEINER (eds.), *Community Participation and Geographic Information Systems*, London, Taylor & Francis, 2002, 17-36.

[SIE 02] R. SIEBER, "Geographic information systems in the environmental movement", in W. CRAIG, T. HARRIS, D. WEINER (eds.), *Community Participation and Geographic Information Systems*, London, Taylor & Francis, 2002, 153-172.

[STO 02] S. STONICH, "Information technologies, public participation GIS, and advocacy: Globalization of resistance to industrial shrimp farming", in W. CRAIG, T. HARRIS, D. WEINER (eds.), *Community Participation and Geographic Information Systems*, London, Taylor & Francis, 2002, 259-269.

[TUL 02] D. TULLOCH, "Environmental NGOs and community access to technology as a force for change", in W. CRAIG, T. HARRIS, D. WEINER (eds.), *Community Participation and Geographic Information Systems*, London, Taylor & Francis, 2002, 192-204.

[VEN 02] S. VENTURA, B. NIEMANN, T. SUTPHIN, R. CHENOWITH, "GIS-enhanced land-use planning", in W. CRAIG, T. HARRIS, D. WEINER (eds.), *Community Participation and Geographic Information Systems*, London, Taylor & Francis, 2002, 113-124.

Chapter 11

Geographic Information and Inter-Organizational Partnerships

11.1. Introduction

Technological evolutions involving geographic information systems (GISs) and the development of spatial databases help in the integration of data between organizations and stimulate inter-organizational partnerships to collect, exchange and use GI in different applications. Spatial data sharing and coordination of GIS development are the foundations of national spatial data infrastructures (NSDIs). NSDIs pertain to technical interoperability and the unrestricted data flows at all levels between administrations, with the private sector and academia. NSDI implementation methods were launched in the USA (*Federal Geographic Data Committee* [FGD 01]), in some European countries [CRA 00, CRA 02, ROC 99, ROC 03, BUR 98], and in other regions of the world [ONS 01].

Infrastructures such as NSDIs make unlimited sharing of spatial data possible and thus avoid duplication of efforts and redundancy in

---

Chapter written by Henri PORNON, Zorica NEDOVIC-BUDIC and Jeffrey PINTO.

the development of geographical databases. However, the rapid development of GISs in organizations highlighted in an interesting way the fact that between and in organizations, it is difficult to share GI and that the level of coordination is low [WAR 98, CAM 95, ECO 94, POR 98a, POR 98b, POR 98c, POR 02b]. These problems are not technical; on the contrary, they come from a wide variety of human considerations explaining why information is still collected and why organizations are apparently resisting the obvious advantages of information sharing [NED 00]. Wasted duplication efforts caused in large part by the lack of exchange of information between administrations and private sector organizations, which remains an important obstacle to a more efficient use of GIS in society and hinders the development and use of the technology [FRA 92, WAR 99, ECO 94]. Among different parameters, the nature and quality of inter-organizational partnerships seem to be determining factors in sharing situations.

GSI technology and inter-organizational partnerships regarding GIS are concerned by the evolution of information technologies (IT) in general. The emergence of open and interoperable systems [BIS 98, LAU 98, CAR 02], distributed data processing and the Internet (*World Wide Web*) for visualization and diffusion of data [HEI 98] resulted in a decentralization of systems and databases. The possibilities offered by Internet technologies in particular are useful in facilitating exchanges, sharing and access to GI. GIS users and other involved parties located in the same territories have begun to use their skills to share the GI and implement the inter-organizational partnerships.

In this chapter, we first address the context of inter-organizational partnerships surrounding GISs. We then present two conceptual contexts used to understand these partnerships, as well as the aspects enabling their creation and operation. Finally, we examine the key success factors. This chapter is based on concrete case studies in France and the USA; the contexts of partnerships may be very different in both countries.

11.2. The context of partnerships

Numerous local and regional organizations create and maintain digital spatial databases to satisfy their individual needs in applications. These public and private organizations located in the same territories are mutually interested in their data and need secure access to (common in particular) data created by the other local organizations. Access and exchange are usually possible with the help of local network operators. Similarly, regional administrations responsible for the environment, water resource and natural risks are interested in GI sharing and cooperation with communities. However, the constitution of databases is often done with no coordination. Consequently, important duplications and data redundancies are the result of these uncoordinated activities [WAR 98], particularly between organizations operating in the same territories. Scales, territories and references involved are variable (topographically at 1/200 in small scale databases), the tools implemented are varied (GIS, CAD, etc.), but all these organizations at one time or another need to access data from other organizations to improve their knowledge of the territory [POR 02b].

The underlying presumption in the development of inter-organizational development for GIS is that they facilitate sharing and exchange of GI for the benefit of the organizations involved. Avoiding useless redundancy of data and duplication of efforts would seem to be the main goals in the initiatives in data coordination. However, the expected advantages of partnerships and sharing go beyond efficiency and include a larger availability of data, better relations between organizations and the diffusion of the technology to smaller organizations, for example, the smaller communities [NED 00, NED 01, POR 02a]. In GIS literature, other reasons to justify GIS partnerships are mentioned: organizational needs and capacity [CAL 95]; relations of power [POR 98a]; professionalism and common goals [OBE 95]; and motivations, goals of the leader, accessibility, as well as shortage of the resource [PIN 95]. [OTO 84] summarizes the different motivations in three categories: authority, common interest and exchange (receiving something in return). It is important to consider these different types of motivations for data exchanges because we expect them to structure the data sharing agreements and partnerships.

The geographical data sharing efforts imply the redefinition of existing tasks and structures and the establishment of new tasks and structures [AZA 95, ECO 94]. An inter-organizational partnership is structured by the definition of roles, obligations, rights, procedures, locations, information flows, data, analysis and computer methods used in the relation [KUM 96]. There are many ways to structure an inter-organizational GIS and database activities. The different configurations in which GIS resources are developed and exchanged depend on institutional, technical and economic constraints [DUE 87, POR 98c] and specific motivations mentioned.

In 1998, Pornon carried out a study for the French Ministry of Equipment [POR 98c]. Based on 20 cases of inter-organizational partnerships, she concluded that the phenomena of power (individual and organizational), implementation costs and problems linked to the technical tools were the main factors limiting these partnerships. The scope of partnerships concerning GI can be identified at several levels, for example: sharing the cost of acquisition and maintenance of reference data (example of the cadastre); communicating available data (metadata exchange); exchanging thematic data on the common territory to improve the understanding of local spatial processes and phenomena and implementing technological solutions for mutual access to data.

[PIN np] uses a scale from the simplest inter-organizational activities to the most complex to describe the intensity of the activities linked to GI in local and regional agencies in the USA. This scale includes the following activities: (1) awareness; (2) communication; (3) exchange of data: receive only, give only, share both ways, share profits, share costs, exchange, no charge; (4) coordinated development of databases; (5) coordinated maintenance of databases; (6) share hardware; (7) share software; (8) share databases; (9) share applications; (10) share personnel; (11) share services or space; (12) share data warehouse.

Several studied partnerships in the USA and France do not go beyond the basic levels. However, in the USA most sharing activities (76.4%) have at least reached a phase where real data exchange is occurring. Even more interesting, almost half (54.4%) of these exchanges are free and one third (33.7%) of exchanges are bidirectional. More complex forms of interaction, such as the common development of a database and sharing resources beyond data (personnel, installations or hardware) is not as widespread. In France, most partnerships focus on sharing costs for external data acquisition (IGN data, Cadastre), but some of them have gone beyond the constitution of common data catalogs or bidirectional data exchange (less than 25%).

However, the conclusions indicate that efforts are made in this field and a considerable number of sharing activities include at least one coordination involving the development of the database. Nevertheless, more important sharing efforts to facilitate direct access to data, such as the development of data warehouses, seem to remain ignored, with only 13% of interactions involving such initiatives in the USA and a few rare experiments in France (Department of Haute-Savoie). This particular aspect confirms the tendency for developing individual GIS identified in the FGDC Framework Study [TUL 01] and shows more trust in less efficient means of access, which could endanger investments in the implementation projects of larger spatial data infrastructures. The North American Framework Study of 1998–1999 showed that ~40% of responding cities and county agencies participated in data coordination committees and that over 80% shared data with other organizations. Most of the respondents involved in GIS data sharing participate in groups of between two and five organizations. Very few organizations (less than 10%) make their metadata or the database itself available [FGD 01, SOM 99, HAR 00, TUL 01] without being concerned with data sharing.

11.3. Conceptual concepts

The systematic understanding of inter-organizational partnerships requires the development and use of conceptual frameworks integrating the components and processes characterizing such associations. Pornon [POR 98c] suggests that inter-organizational partnerships could be represented by a conceptual framework that includes: on the one hand, the state reached by the partnership from the viewpoint of computer tools, data, degree of legal formalization reached and quality of the relation between the stakeholders in the partnership (static aspects of the partnership); and on the other hand, the evolution of the partnership (dynamic aspects). [NED 99a] uses another framework that involves four elements: context (organizational and inter-organizational); motivation; mechanisms of coordination (structure, processes and rules); and results. The two frameworks are presented below.

11.3.1. *Framework proposed by H. Pornon*

11.3.1.1. *Static aspects of the partnership*

Partnerships can be described by a large number of parameters. Considering that strategic aspects are the major problems, H. Pornon chose to concentrate on four of them (Figure 11.1). Two technical parameters identify the level of partnership reached by the participants: the computer tools used and the data involved. Two other parameters qualify the relations between partners: the level of formalization (legal structure) shows what arrangement was found and the "humanware" illustrates the quality of relations between the participants (individuals). The financial aspect was not taken into consideration, as financial arrangements are considered as a consequence of the other aspects. Each aspect is defined as a scale in which the first level is equal to 0 (expressing the lack of partnership); each higher level indicates a higher level of mutual trust and partnership activity.

The arrows between the boxes suggest dependence between the aspects involved. The significance of the levels is described in Tables 11.1, 11.2 and 11.3. The scale in the "data" box is based on the 20 cases studied by [POR 98c]. He observes that the partnerships often start by sharing costs of data acquisitions (e.g. digitization of the cadastre). The following steps, involving a higher level of mutual trust, are metadata sharing (constitution of common catalogs), data exchanges, and sometimes, the coordinated or common management of data between partners. The highest level of association involves common data management.

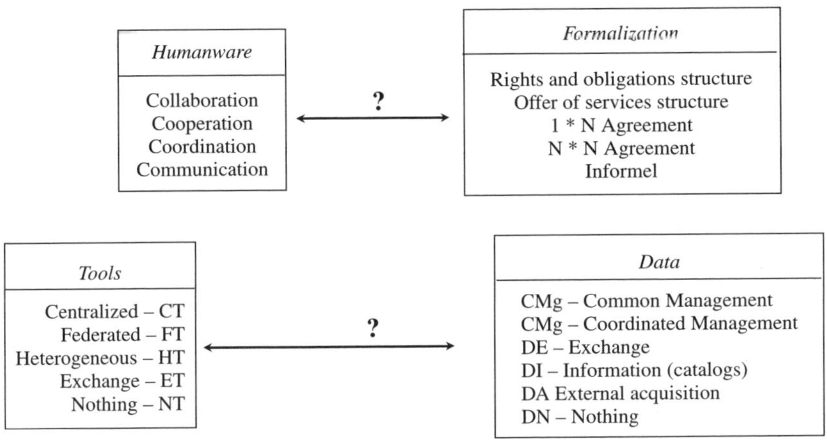

Figure 11.1. *Static aspects of the partnership*

The scale involving computer tools was defined in Pornon's thesis [POR 98a]: it has a typology of GIS architectures based on the autonomy/coordination dialectic, which applies to partnerships studied as well as to the description of intra-organizational GIS.

The first level of coordination is the use of exchange formats and protocols: each organization has its own GIS and data exchange on demand by using defined formats and protocols. In the second level, a shared server is used, but each partner can use its own GIS using gateways to access the common server. In the federated architecture, all the organizations use the same GIS and their data is centralized on a server. This solution is mostly used between local councils and small villages. The most centralized solution is the one in which one organization manages the GIS for all partners. Emphasizing the advantages of centralized processing (and contradicting the arguments supporting distributed processing), [JAI 98] concludes in his study on data resource management that a "large centralization of decisions concerning information systems, a large concentration of resources in a central site, and low autonomy granted to local sites" is a guarantee of success for the system. This point of view is contradicted by the problem in making partnerships work based on such technical solutions.

Type	Effect
Common management	Some partners accept that their data be managed by another partner.
Coordinated management	Partners make their data available to other stakeholders and accept that they influence the character of the data.
Data exchange	Partners open their data to other stakeholders.
Data catalog	Each partner described his data, but does not open his data to other stakeholders yet.
External acquisition	In the case of shared cost external data acquisition, data from the organizations is not involved.
Nothing	No partnership between organizations involving data.

Table 11.1. *Data and partnership*

The human factor is an important part of this model, considering that the mutual trust between individuals representing their organizations can determine the quality and operation of the partnership. The "humanware"[1] factor is based on the adaptation of a typology suggested by [JAN 97], and excluding the level 0 (no relation):

– The first level is [Communication]: the stakeholders talk to each other (but do not yet have to be concerned about harmonizing their actions). Example: reciprocal information on the mutual projects of partners, prior discussions to a partnership project.

1 The choice of this term comes from the necessity in finding a qualifier other than the four terms used to describe the levels of the scale.

– The second level is [Coordination]: the stakeholders have to harmonize their actions together, but not between each other. Example: sharing results of studies, investigations, services, exchange of recipes, methods, tips, expertise, data sharing, etc. The acquisition of shared cost data or data exchange with no effort in their making sense seems to us to belong in this category.

– The third level is [Cooperation]: the stakeholders are involved in common accomplishments, but do not necessarily have the will to cooperate beyond the expected accomplishment: they share common tasks, but not objectives. Example: sharing the entry and updating of data between partners. Since the development of common data catalogs and data exchange between partners requires cooperation, an effort in consistency or some sort of division of responsibilities is part of this category.

	Coordination	Autonomy
CT	Centralized system: an organization proposes to be application supplier for the others	No GIS in the other organizations
FT	Federating system: Inter-organizational server in which all organizations share the same GIS	Data from each organization is stored in a centralized database on the server
HT	Heterogeneous client/server system: Inter-organizational server with gateway to GISs of organizations	Autonomous GIS in organizations, periodically exchanging data in both directions with the server
ET	Data exchange: definition of exchange protocols between organizations: format, support…	Autonomous GIS in organisations, exchanging data on demand between each other
NT	No coordination	No exchange between GISs in organizations

Table 11.2. *Computer tools and partnership*

– The fourth and highest level is [Collaboration]: the stakeholders participate in common operations, by doing common tasks on the basis of objectives exceeding these accomplishments (we pursue common objectives involving the territory). Examples: implementing common projects to enhance common know-how, using a common GIS to make decisions together.

Formalization	Comments
Rights and Obligations Structure	Legal structure offering services, but where clients also have ogligations
Service Offer Structure	A legal structure offering services but where clients do not have obligations
1 * N Agreement	Partnership agreements in which an organization plays a preferred role
N * N Agreement	Bilateral partnership agreements between organizations
Informal	No formalization

Table 11.3. *Partnership formalization*

The last scale involves formalization, or in other words, the legal structure of the partnership (Table 11.3). Beyond the informal phase, inter-organizational partners are often legally represented by one or more agreements or contracts between partners or by the creation of new legal structures. The agreements provide more structure if a partner plays a preeminent role (1*N type convention, as opposed to N*N type bilateral agreements). The other formal structures are more restrictive, particularly when the partners have not only rights (for example, access to a database) but also obligations (for example putting their own data in a shared database).

11.3.1.2. *Dynamic aspects of partnerships*

The static model boxes show the steps reached by the partnership, notably the current phase. However, if we concentrate on the evolution of partnership, or if we wish to characterize the itineraries of the partnerships, we need a dynamic representation of the partnership over time (Figure 11.2). The question raised in particular is to know how the formalization and "humanware" evolve with tools and data. Within each box, represented in the "humanware" and "formalization" diagram, the names of organizational leaders are indicated (for example, CETE in the beginning and CR as in Regional Council), as well as the phases reached for data and computer tools (Figure 11.2).

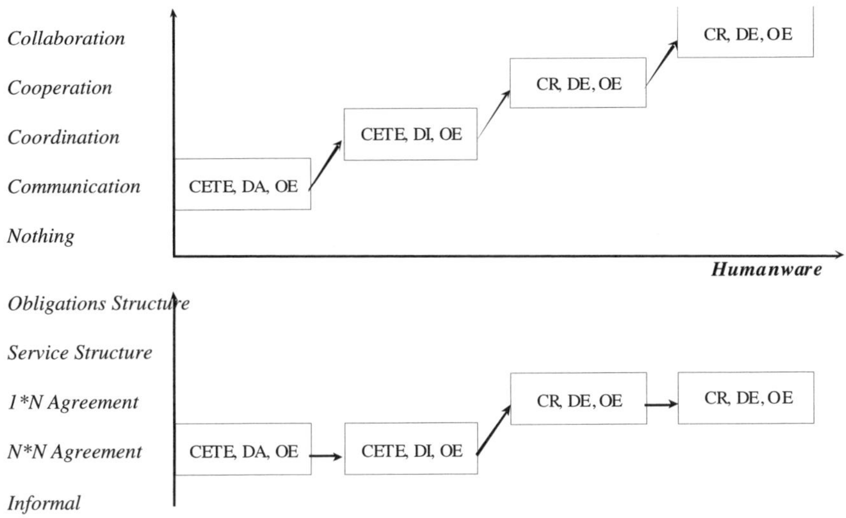

Figure 11.2. *Itinerary of partnerships*

Based on the 20 cases, [POR 98c] identifies two approaches (Figure 11.3): "priority to formalization" is the case where the initiator of the partnership focuses on the legal aspects and starts with the creation of a structure, enabling him to become the leader of the association. His problem is generally to convince potential partners and create trust with and between them.

In the "priority to humanware", the stakeholders start by discussing and try to create mutual trust. They remain informal as long as possible to avoid potential conflicts that could surface from the formal framework. The agreement from this point of view is a more flexible solution, but usually there comes a time when they have to move into formalization.

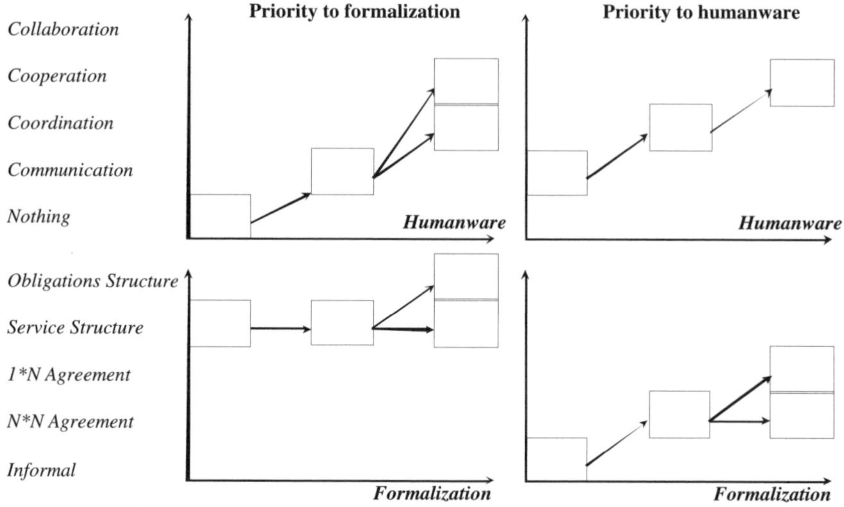

Figure 11.3. *Itinerary of partnerships*

In many projects that Pornon studied, the steps are similar (Figure 11.4). The partnership starts easy enough by sharing the cost of acquiring an external database. Except for the definition of common formats, no computer tool is required and data from partners is not involved. This enables partners to go from communication to coordination, and partnership agreements are enough. No structure is necessary.

Participating organizations then often start to question their data and develop a common data catalog. Sometimes, they discuss more ambitious projects, such as heterogenous servers (Internet or spatial server). Finally, when they start sharing goals, they can exchange data and share tools.

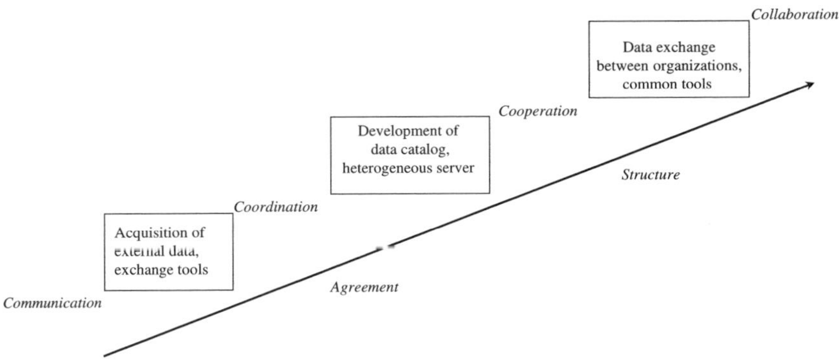

Figure 11.4. *A common itinerary*

11.3.2. *The framework proposed by Z. Nedovic-Budic and J. K. Pinto*

The literature on inter-dependencies between organizations, internal and environmental parameters of organizations, inter-organizational partnerships and the implementation of policies provides several concepts and paradigms useful to guide the studies and management of common development and geographical data sharing. [KEV 95] provides a complete list of relevant parameters of GIS data sharing from his professional experience. These parameters include the following: sharing of categories; organizational environment; the need for shared data; the opportunity of sharing data; the will to share data; the motivation to share data; obstacles to share; technical capacity to share; and resources to share.

The conceptual framework presented here is based on Kevany's ideas, but refers to a more general literature to bring out four general theoretic constructions of the coordinated GIS development and database sharing. These constructions include the following: context; motivation; coordination mechanisms (structure, process and policies); and results (Figure 11.5).

The inter-organizational context refers to the important organizational parameters for the coordination and decisions involving common GIS. Despite the limited capacity in the manipulation of contextual factors, and particularly considering the future scale of an NSDI, the context is very important to understand the partnerships concerning GI.

Inter-organizational database systems are the representation of relationships between organizations [KUM 96] and administration models [WES 91]. Associating the context and GIS activities should help in evaluating the applicability and functions of sharing structures and policies in the different inter-organizational frameworks and in different organizational situations. [AZA 98] maintains that "working the context" is necessary for improving the chances of success of an organizational GIS.

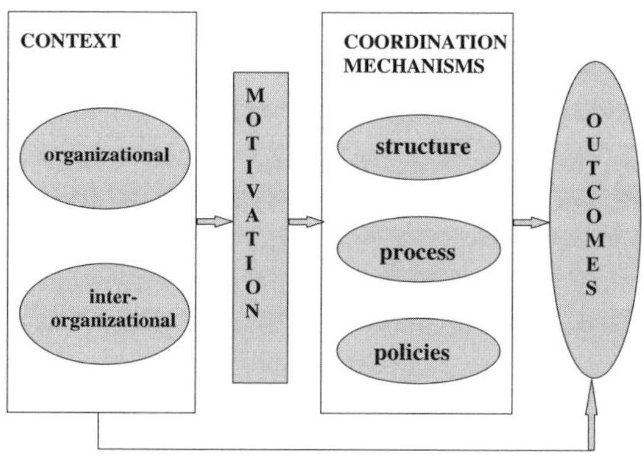

Figure 11.5. *Conceptual framework of Z. Nedovic-Budic and J. K. Pinto*

The need to identify the motivations of organizations to actively get involved in relationships with other organizations are fundamental to the discussion on the interest of inter-organizational GIS coordination. Several factors contribute to the perception of the need of relations between organizations around GI. The coordination can be imposed by the authority, i.e. stemming from a sense of duty; by the common interest of organizations sharing the same common goals; or still by the exchange of services when a return is expected or received. Participation can be intentional or imposed [CUM 80].

The importance and content of inter-organizational interaction are often determined by the main function of one of the organizations [LEV 69]. In GIS literature, the following reasons are mentioned to motivate their interaction: organizational needs, capacity and cost [CAL 95]; relations of power [POR 98a,c]; professionalism and common goals [OBE 95]; motivations, goals of the main leader, accessibility, as well as shortage of the resource [PIN 95].

The main element of the conceptual framework presented is based on the partnership involving GI. This partnership integrates the inter-organizational structure, the policies used, history and processes involved in the coordination of multi-participant GIS and in the establishment of sharing relations.

Identifying and understanding more robust coordination mechanisms and the most efficient policies can reveal to be of great use to develop partnerships in terms of GI. The establishment of formal standards, agreements and coordination structures for example constitute the preconditions to any distributed GIS [MER 95]. Formalization tends to facilitate institutionalization of common behaviors, grouping and commitment [BRO 98].

Finally, the direct effects of inter-organizational GIS and database activities in specific fields of application constitute the first tangible results and can be considered as a measure of the value and social usefulness of GI [ONS 95] (it is another way of expressing the role of the useful value (see Chapter 5)). The advantages from GIS sharing cannot be distinguished from the value of final decisions and the social effects to which they contribute. In theory, organizational partnerships involving GI seem to offer significant advantages in:

– efficiency (costs avoided and gains in productivity in existing operations at a lower cost per unit);

– effectiveness (improves capabilities of units; new products of better quality; improved policy and better decisions);

– benefits for the company (new responsibilities; generalized mission);

– public service [BRO 93, DUE 95, DAW 96].

Based on the conclusions from several cases studies, [NED 00] revised this conceptual framework to define in a more explicit way the structures and processes of inter-organizational partnerships (Figure 11.6). It is important to note that this framework is a multi-dimensional concept. The results from case studies showed that five aspects of this framework have a particular importance in studying interaction mechanisms [NED 99a,b, NED 00].

First, we can identify several forms of sharing mechanisms based on the scope of the interaction involved. This scope can vary from a simple awareness or communication concerning GIS activities to different forms of data exchange and, in its most advanced manifestation, integrate the development and/or system maintenance and resource sharing other than data (personnel, installations and hardware).

Second, interaction mechanisms can be formal or informal.

Third, different standardization activities can be involved, going from no acceptance of standards to the adoption of private, local, regional, federal or international standards.

Fourth, an association can be characterized by the status of participation of the entities involved. Different forms of adhesion with different levels of influence or simple execution statuses can be distinguished.

Finally, the structure of partnership relation can be characterized based on the contributions of entities to the development of the common GIS databases or to activities of GI exchange.

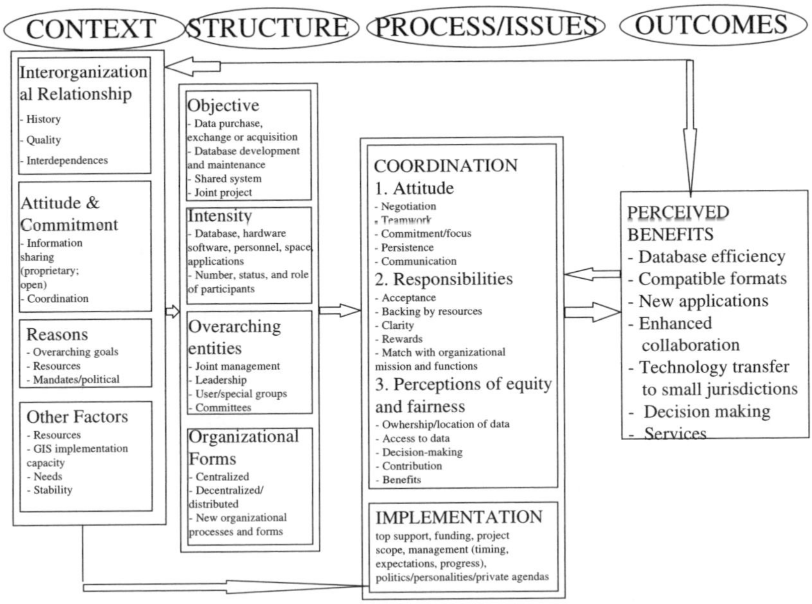

Figure 11.6. *Extended conceptual framework of Z. Nedovic-Budic and J. K. Pinto*

11.4. Other aspects involving GI sharing

11.4.1. *Legal aspects*

It may be useful at this point to recall a few legal constraints involving geographic data, where the impact is particularly important in the case of Internet applications (which constitute the vital medium of GI sharing today). The first one involves the respect of the data owner's prerogatives, even in the case where the data is public (in France, IGN or Cadastre data, etc.). We do not have the right to broadcast IGN data, even as background, without consent (thus in practice without paying royalties). The second one involves the civil and penal responsibility of the broadcaster, linked to the diffusion of data (the case of a city that would want to broadcast data on the networks of its claim holders for example). The risk is great in case of an accident. Finally, the respect of the CNIL measures [Commission Nationale Informatique et Libertés (National Information Technology and Freedom Commission)] is the third main constraint: there is sensitive data and generally, it is better to avoid publishing nominative data over the Internet.

In the USA, all levels of administration use GIS technology in different applications, including public safety and security, design and management of infrastructures, cadastre, urban planning, natural resource and environment management and emergency services. As in other fields, activities linked to GIS are concerned by the federal legal and constitutional environment: the first amendment to the American Constitution guaranteeing freedom of expression and freedom of the press; Federal Freedom of Information Act (1996, 1974) mandating free access to the public of data; the Copyright Act (1976); the Privacy Act (1974, 1988); and intellectual property, responsibility and antitrust law [JOH 95]. The three levels of administration produce public domain data, but the mandates for the diffusion of data collected or developed with the revenue department are first attributed to the agencies of the federal government. There is also guaranteed access to public data collected and maintained locally. Interestingly, despite the common legal environment, the States and House of Commons interpret the right to access digital data differently. Case studies from [JOH 95] show how the policy of access to information varies between the counties under the same law concerning the public registry and even to laws of the State. In North Carolina for example, some counties provide open access to digital GI; others want to recover "reasonable cost" and impose restrictions to access, authorized by an Act passed in 1991 by the General Assembly of North Carolina: "Act representing a Qualified Exception to the Act of Public Registry for Certain Geographic Information Systems".

11.4.2. *Standardization and interoperability questions*

Another series of problems in terms of information sharing, not identified as often but just as important, involves semantic questions [KUH 03]. These problems are not specifically linked to the Internet;

they involve all inter-organizational partnerships, regardless of technological support. These problems concern the relevance of shared information in particular: everyone does not work on the same scale and is not interested in the same themes, or the same spatial problems. The variety of scales, representations and specifications is often an obstacle for reusing information from one organization by another: a departmental or regional administration will not see the monuments or roads the same way as a village (point/surface, line/surface). The outline of villages extracted from the cadastral map is too detailed for regional use. The village may be interested by a zoning map rigorously targeting lots, administrations responsible for urban planning by zoning maps form a partition of the departmental or regional territory: these specifications are contradictory. Information sharing or data exchange first presumes metadata communication.

In general, Webster's Collective Dictionary ([MER 93], 10th edition) defines interoperability as the "capacity of a system (as a weapons system) to use the parts or material of another system". Except for the military industry, interoperability has become a common term among the practitioners and researchers in the field of GIS. The term takes a variety of definitions from software and hardware to formats and models of spatial data [ALB 99] and to semantics [HAR 99]. [BIS 98] distinguishes two levels of interoperability: application semantics, data models, database management systems (DBMS), spatial data files, hardware and operating system (OS) and network protocols.

These levels of interoperability are used to develop the notion of geospatial information community (GIC) or groups of users sharing a vocabulary and definitions of common objects [BIS 98, 99]. The definition of GIC implies data sharing but is also based on concepts which do not include organizational environments and parameters. [NED 01] introduced the term "soft interoperability" to attract attention to human and organizational aspects of GIS and database sharing between organizations.

In the field of GI, the standards are particularly important and necessary [GUP 94]. The goals of the standards consist of reducing the obstacles to interoperability. Data interoperability is defined as "the capacity of accessing environments of analysis and multiple and heterogenous spatio-temporal processes, local or remote with the help of a single software interface" [BUE 96]. Using standards can also reduce the cost of data sharing [BOS 95] and helps in the determination of data appropriateness to different user needs. Inter-organizational information systems require standards at different levels of development, processing and use of GI: the data model, georeferencing, types of spatial data, content of specific data layers, data collection procedures, quality of data sources, precision of data, metadata, production specifications, transfer and use of data, access to databases, catalogs of available data [KUH 03, LAC 03], data format and the design of the database [NED 99b].

11.4.3. *Sociological and organizational aspects*

Another dimension of the problem raised by information sharing is sociological. Competitive situations between people or services in a single organization or between organizations create barriers to information sharing: between Administrations and communities, between Villages and Local Councils, between Communities and Developers, between Administrations or between companies, etc. We should not ignore these phenomena of power, which are the main problem with inter-organizational partnerships [POR 98a,c]. The Internet technologies cannot solve these problems, obviously, but they provide opportunities for evolving the situations. In fact, it is easier to "avoid" solicitations of exchange when they cause technical problems and lead to important costs (pre-2000 situations) than when there are simple formats, standard functions and inexpensive tools facilitating exchanges (current situation).

The general implementation of GIS, particularly in GI sharing situations, is a problem of organization dominated by inter-organizational structures [AZA 95, POR 98c]. First, implementing a database or a common system increases interdependence between organizations. By participating in this type of partnership, organizations must lose some of their autonomy. [AZA 95] highlighted that the scope with which the autonomy of organizations is affected determines the probability of success of the partnership. To avoid the negative effects of inter-organizational projects and maximize their organizational gains, agency civil servants have a tendency to form partnerships liable to affect their autonomy as little as possible and to minimize their loss of power [COO 77]. Consequently, the development of common activities brings its own problems. The argument of interdependence and the need for cooperation can often lead to radical changes in situations [STE 71], power struggles [DAW 96] and to conflicts of authority, jurisdiction and distribution of power [KUM 96, POR 98a,c].

Inter-organizational exchanges are affected by different parameters of organization and environment [CUM 80] such as environmental stability, homogeneity of organizations, concentration and capacity of resources, the consensual field between organizations, disruptions and changes and finally, the capacity for change of organizations in their environment. Finally, coordinated activities between organizations, linked to GI not only depend on organizational and environmental parameters but also involve organizational, sometimes radical, changes [NED 99c].

11.5. Key success factors of inter-organizational partnerships

11.5.1. *Aspects favoring or hindering the development of partnerships*

The maturity of Internet technologies is not the main success factor for inter-organizational partnership projects. Collaborative work habits of organizations are certainly the most important parameter. Case studies carried out by [POR 98c] show that in cases where there were cooperation situations between communities and administrations, before the implementation of inter-organizational GIS partnerships (the case of the Nord-Pas de Calais region, for example), partnerships progress more quickly. This aspect is confirmed by the study of several Missions Inter-Service de l'Eau (MISE – Inter-Service Water Mission) in France. These MISE are an example of cooperation between more than 10 local administrations managing together the water police on rivers and water resources and attempt to develop complementary databases together. Some work as real workgroups, make decisions together and use collaborative computer tools and common data. Others are still in the phase of communication (mutually communicating the decisions made individually by each organization) and are not considering the deployment of common tools.

In a more general way, even though Internet technologies offer interesting opportunities, mobilization of a few key stakeholders, focusing projects on negotiation and the human dimension of projects instead of on technology, the identification of strategic situations (important role of MISE in the water law for example) and the perception of challenges by policies are the real success factors for projects.

Reciprocally, hindrances to the development of Internet technologies for partnership are rarely technological in nature. Situations of competition between organizations, the low perception of stakes by decision makers, the high cost of coordination, tools and shared geographical data, turnover of the participants (we must sometimes start from scratch) and strategic and sociological phenomena are the real constraints.

11.5.2. *Advice to increase the chances of success in projects*

11.5.2.1. *Keep it simple*

The scope of interactions between organizations usually goes beyond the aspects linked to data (acquisition of data; data exchanges; projects involving common data; common acquisition of data; development and or management of common databases), to include systems developed in common, personnel (often in the role of coordination and technical assistance), locations and applications. The last aspect is most often relevant in an intra-organizational context. From data to applications, interactions increase in sophistication and complexity, but also in the difficulty in making them operational. Joint support or sharing of the development of applications is the trickiest aspect. We suggest to think big but start small and gradually build the partnership around approaches centered on data.

11.5.2.2. *Formalize and structure*

Mutual trust is the key to prosperous cooperation, but it would be wise to prepare with formal and legal documents, such as inter-administration agreements, memorandum, licenses relating to data, contracts, etc. They are more commonly used (and probably more important) in relations with other organizations than within a single organization. These documents ensure continuous data exchange even when other forms of interactions are interrupted. The nature of sharing structures also needs to be established early on in the process. Allowing interactions between GIS and databases to evolve simply over time without rules and procedures often attracts dubious partners and can lead to real problems. Consequently, the key is to establish a stable, simple and permanent relation structure.

11.5.2.3. *Contributions: justice, equity and continuity*

It is first necessary to determine the contributions in advance and in specific terms. Data is the most important contribution to coordinated activities. Financial and personnel contributions are also significant; the latter is more obvious in an intra-organizational context. The concerns of most organizations involving sizing of their contribution in relation to their size, resources, the use of their data or other common products are all aspects that need to be considered. The principle of equity in the acceptance of contributions and the distribution of common resources must be applied. Long negotiations can be necessary to define contributions and recipes. The lack of independence and the investment in energy and resources are real and tangible contributions in the development and maintenance of relations with other organizations. A minimum level of contribution from each participant tends to increase the adhesion to common goals and improves the chances of success. Finally, we must ensure long-term commitment of contributions. Their variations from one year to the other can endanger the project and prevent the implementation of strategic or more important pluriannual projects.

11.5.2.4. *Control and propriety*

As in many inter-organizational projects, participants need to feel authorized to organize, make and implement decisions. The participants of inter-organizational activities rely on a process of honest decision, to ensure adequate control of common activities and resources. Rights to vote and decisional authority must be safely and clearly defined. Long negotiations can also be necessary on this point. We must expect that the viewpoints of organizations on equity and justice in the representation of decision processes are different according to their resources, their power and role in the association. Perceptions involving the property of data contribute essentially to these verifications. The opening in data access, the minimum propriety interest on data and the lack of major financial gains expected in their diffusion limit the conflicts and tensions concerning the property of data. Finally, all the parties must perceive the coordinator and his position as neutral (i.e., with no capital invested or no commitment in any agency or organization).

11.5.2.5. *The "what is in it for me ?" syndrome*

The "what is in it for me?" question must be taken seriously. The understanding and respect of reasons motivating the participation of organizations are a part of the success. The economy of resources, missions and common goals are the most frequently used reasons to justify interactions between organizations.

11.5.2.6. *Act, act, act*

Streamline communication and negotiation are two aspects essential to efforts of coordination. In fact, the identification of semantic differences and similarities between the concepts used by participants, as well as the creation of a common working vocabulary are the preconditions of efficient communication. Communication can be formal and/or informal. The negotiation to build a "coalition" must be carried out carefully. Tenacity and good will in finding compromises are keys to success, especially in difficult moments (which exist even in the most advanced partnerships). Different levels of commitment can be considered. A real commitment helps in overcoming many obstacles in database implementation processes or common system also helps in maintaining the priority direction of common activities. Participants who have committed "for the wrong reasons" are usually resistant to common effort. The process takes time and patience. Finally, the spirit of cooperation is crucial to keep participants active and interested. It is based on teamwork, shared understanding, mutual trust and credibility.

11.5.2.7. *Project leadership*

It is the key factor of success. The leader provides vision, support and resource backing. He has the authority to act quickly on common methodologies and decentralizes it for implementing parts of the project specific to agencies. The stability characterizes efficient project direction structures. The leader ensures that common standards are being used and commitments are executed (for example, databases or other contributions).

11.5.2.8. *Roles and responsibilities*

Roles and responsibilities of each participant must be perfectly defined. It is another factor of success. The development of the database and responsibility of its maintenance are the lifeline of partnerships. It is necessary to identify and to acquire the support of basic data providers early on in the initiative of coordination, for the collection and update of data be achieved close to the source and/or in organizations with adequate tools. Additional resources and the support infrastructure must be available to organizations, as well as the new roles and responsibilities leading to additional work and expenses (for example, responsibility for updating data).

Organizations perceiving injustices in commitments involving data updates may want to decrease their own participation in the system. In the absence of personnel recruiting, financing, hardware or training, agencies ensuring responsibilities of database maintenance may experience delays and deteriorations in the quality of database updates. They also have a tendency to distance themselves from the stipulated standards. The assignment of roles and responsibilities is very sensitive to questions of equity.

11.5.2.9. *Management of change*

In a very technical field such as GIS, it is necessary to adapt local situations to benefit from technological changes and innovations. The disparity problems between the new tasks of the database and procedures and structures of existing organization are common in recent partnerships. The technological change requires a change in administrative and organizational structures and processes. Integrated and distributed data processes tend to produce "leaner" organizations, more flexible and reactive with less levels of management and more direct information exchanges between the high and low end of the hierarchy.

The perception of the future change and the uncertainty that it brings have a tendency to disrupt many organizations as well as their personnel. It is crucial to face the concerns involving the implications of technological change and common activities of the database on the organization and the personnel. The state of advancement of the common project must often be presented and communicated to all participants and people responsible. The situation of the project should be managed at administrative, managerial and operational level. Finally, the culture of sharing and change must be maintained.

11.5.3. *Lessons learned from the [JOH 01] study*

A few additional lessons can be learned from a study of six partnerships carried out in the USA by [JOH 01]. These partnerships were created with different objectives: information sharing forum, geospatial data publication sharing forum, execution of integrating functions involving data from multiple sources in a region and a combination of these objects with others. The study identifies 17 fundamental practices in their capacity to reach their respective visions. Below are six key aspects shared by all participants (they are cited with no order of priority):

– general support of visions and expectations,

– support of individuals/support of the community,

– well informed participants taken into consideration,

– maintenance of contact with national organizations (of higher rank),

– proactive process, open and comprehensive/procedures encouraging maximum participation/different aspects,

– promotion of the agreement and results.

In addition, five additional keys to success were cited by four of the six participants. They are cited below with no specific order or priority:

– support of organizations,

– diffusion of benefits for parties involved and of economic arguments,

– maintenance of the focus of common information needs,

– maintenance of institutional memory,

– support of the business plan/well-defined aspects.

11.6. Conclusion

These different studies carried out in the USA and in France converge toward the same observation: even if new information technologies, notably the Internet, offer very interesting opportunities to federate GI and enable its sharing between organizations, the problems are many and are mainly involved with phenomena of power and political and organizational contexts. The models proposed in this chapter to support the understanding of these partnerships also provide action methods to increase their chances of success.

11.7. Bibliography

[ALB 99] J. ALBRECHT, "Towards interoperable geo-information standards: A comparison of reference models for geo-spatial information", *The Annals of Regional Science*, vol. 33, 151-169, 1999.

[AZA 95] B. AZAD, L. WIGGINS, "Dynamics of inter-organizational geographic data sharing: A conceptual framework for research", in H.J. ONSRUD, G. RUSHTON (eds.), *Sharing Geographic Information,* New Brunswick, NJ, Center for Urban Policy Research, 1995, 22-43.

[AZA 98] B. AZAD, *Management of Enterprise-wide GIS Implementation: Lessons from Exploration of Five Case Studies*, MIT Press, 1998.

[BIE 01] K.T. VAN BIESEN, *Models of National GI Association in Europe,* EUROGI, available at: http://www.eurogi.org, 2001.

[BIS 98] Y.A. BISHR, "Overcoming semantic and other barriers to GIS interoperability", *International Journal of Geographic Information Science*, vol. 12, 299-314, 1998.

[BIS 99] Y.A. BISHR, H. PUNDT, W. KUHN, M. RDWAN, "Probing the concepts of information communities – A first step toward semantic interoperability", in M. GOODCHILD, M. EGENHOFER, R. FEGEAS, C. KOTTMAN (eds.), *Interoperating Geographic Information Systems*, Norwell, MA, Kluwer, 1999, 165-180.

[BRO 93] M.M. BROWN, J.L. BRUDNEY, Modes of geographical information system adoption in public organizations: Examining the effects of different implementation structures, *Annual Meeting of the American Society for Public Administration*, 1993.

[BRO 98] BROWN M. M., O'TOOLE L.J. Jr. and BRUDNEY J.L., "Implementing Information Technology in Government: An Empirical Assessment of the Role of Local Partnerships". *Journal of Public Administration Research and Theory*, 1998.

[BUE 96] K. BUEHLER, L. MCKEE, *The Open GIS Guide,* Wayland, MA, The Open GIS Consortium, Inc., available at: http://www.ogis.org/guide/guide1.htm, 1996.

[BUR 98] P. BURROUGH, I. MASSER, *European Geographic Information Infrastructures,* London, Taylor &Francis, 1998.

[CAL 95] H.W. CALKINS, R. WEATHERBE, "Taxonomy of spatial data sharing", in H.J. Onsrud, G. Rushton (eds.), *Sharing Geographic Information*, New Brunswick, NJ, Center for Urban Policy Research, 1995, 65-75.

[CAM 95] H. CAMPBELL, I. MASSER, *GIS and Organizations: How GIS are Effective in Practice*, London, Taylor & Francis, 1995.

[CAR 02] R. CARROL, "Enterprise spatial data sharing using OPENGIS standards", *Proceedings from URISA 2002*, Chicago, 2002.

[COO 77] K.S. COOK, "Exchange and power in networks of interorganizational relations", *Sociological Quarterly*, vol. 18, 62-82, 1997.

[CRA 00] M. CRAGLIA, P. SIGNORETTA, "From global to local: The development of local geographic information strategies in the United Kingdom," *Environment and Planning B,* vol. 27, 777-788, 2000.

[CRA 02] M. CRAGLIA, A. ANNONI, R. SMITH, P. SMITS (eds.), *Spatial Data Infrastructures: Country Reports 2002*, European Commission, 2002.

[CUM 80] T.G. CUMMINGS, "Interorganization theory and organizational development", in T.G. CUMMINGS (ed.), *Systems Theory for Organization Development*, New York, Wiley, 1980.

[DAW 96] S.S. DAWES, "Interagency information sharing: Expected benefits, manageable risks", *Journal of Policy Analysis and Management*, vol. 15, 377-394, 1996.

[DID 90] M. DIDIER, *Utilité et Valeur de l'Information Géographique*, Paris, Economica, 1990.

[DUE 87] K.-J. DUEKER, "Multipurpose land information systems: Technical, economic, and institutional issues", *Photogrammetric Engineering and Remote Sensing*, vol. 53, 1361-1365, 1987.

[DUE 95] K.J. DUEKER, R. VRANA, "Systems integration: A reason and a means for data sharing", in H.J. ONSRUD, G. RUSHTON (eds.), *Sharing Geographic Information*, New Brunswick, NJ, Center for Urban Policy Research, 1995, 149-171.

[ECO 94] C. ECOBICHON, *L'information Géographique: Nouvelles Techniques, Nouvelles Pratiques*, Paris, Hermès, 1994.

[FGD 01] FEDERAL GEOGRAPHIC DATA COMMITTEE (FGDC), available at: http://www.fgdc.gov, 2001.

[FRA 92] A.U. FRANK, "Telecommunication and GIS: Opportunities and challenges", in P.W. NEWTON, P.R. ZWART, M.E. CAVILL (eds.), *Networking Spatial Information Systems*, London, Belhaven, 1992, 235-250.

[GUP 94] S.C. GUPTILL, "Spatial data standards", *Government Information Quarterly*, vol. 11, 387-401, 1994.

[HAR 99] F. HARVEY, W. KUHN, H. PUNDT, Y. BISHR, C. RIEDEMANN, "Semantic interopérabilité: A central issue for sharing geographic information", *The Annals of Regional Science*, vol. 33, 213-232, 1999.

[HAR 00] F. HARVEY, *Potentials and Pitfalls for Vertical Integration for the NSDI,* Final Report of a Survey of Local Government Perspectives, available at: http://www.uky.edu/AS/Geography/ppnsdi/reports/reports.htm, 2000.

[HAR 01a] F. HARVEY, "Constructing GIS: Actor networks of collaboration", *URISA Journal*, vol. 13, 29-37, 2001.

[HAR 01b] F. HARVEY, "NSDI from the trenches – Local government perspective", *Geospatial Solutions,* vol. 11, 2-4, 2001.

[HEI 98] E. HEIKKILA, "GIS is dead; Long live GIS!", *Journal of the American Planning Association,* vol. 64, 350-360, 1998.

[JAI 98] H. JAIN, K. RAMAMURTHY, H.S. RYU, M. YASAI-ARDEKANI, "Success of data resource management in distributed environments: An empirical investigation", *MIS Quarterly,* vol. 22, 1-29, 1998.

[JAN 97] P. JANKOWSKI, T. NYERGES, A. SMITH, T. MOORE, E. HORVATH, "Spatial group choice: A SDSS tool for collaborative spatial decision making", *IJGIS*, vol. 11:6, pp. 577-602, 1997.

[JOH 95] J.P. JOHNSON, "Case studies of dissemination policy in local government GIS agencies", *Computers, Environment and Urban Systems,* vol. 19, 373-390, 1995.

[JOH 01] R. JOHNSON, Z. NEDOVIC-BUDIC, K. COVERT, J.K. PINTO, *Lessons from Practice – A Guidebook to Organizing and Sustaining Geodata Collaboratives,* Reston, VA, GeoData Alliance, 2001.

[KEV 95] M.J. KEVANY, "A proposed structure for observing data sharing", in H.J. ONSRUD, G. RUSHTON (eds.), *Sharing Geographic Information*, New Brunswick, NJ, Center for Urban Policy Research, 1995, 76-100.

[KUH 03] W. KUHN, M. RAUBAL, "Implementing semantic reference systems," *Proceedings from AGILE 2003*, Lyon, 2003.

[KUM 96] K. KUMAR, H.G. VAN DISSEL, "Sustainable collaboration: Managing conflict and cooperation in interorganizational systems", *MIS Quarterly,* vol. 20, 279-300, 1996.

[LAC 03] J. LACASTA, J. NOGUERAS-ISO, M.P. TORRES, F.J. ZARAGA-SORIA, "Toward the geographic metadata standard interoperability", *Proceedings from AGILE 2003*, Lyon, 2003.

[LAU 98] R. LAURINI, "Spatial multi-database topological continuity and indexing: A step towards seamless GIS data interoperability", *International Journal of Geographical Information Science,* vol. 12, 373-402, 1998.

[LEV 69] S. LEVINE, P.E. WHITE, "Exchange as a conceptual framework for the study of interorganizational relationships", in A. ETZIONI (ed.), *A Sociological Reader on Complex Organizations*, 2nd Edn., New York, Holt, Rinehart & Winston, 1969.

[MAS 98] I. MASSER, *Governments and Geographic Information,* London, Taylor & Francis, 1998.

[MER 93] MERRIAM WEBSTER'S COLLEGIATE DICTIONARY, 10th Edn., Springfield, MA, Merriam-Webster Inc., 1993.

[MER 95] P.-H. MEREDITH, "Distributed GIS: If its time is now, why is it resisted?", in H.J. ONSRUD, G. RUSHTON (eds.), *Sharing Geographic Information,* New Brunswick, NJ, Center for Urban Policy Research, 1995, 7-21.

[NED 99a] Z. NEDOVIC-BUDIC, J.K. PINTO, "Understanding interorganizational GIS activities: A conceptual framework", *Journal of the Urban and Regional Information System Association (URISA)*, vol. 11, 53-64, 1999.

[NED 99b] Z. NEDOVIC-BUDIC, J.K. PINTO, "Interorganizational GIS: Issues and prospects", *Annals of Regional Science*, vol. 33, 183-195, 1999.

[NED 99c] Z. NEDOVIC-BUDIC, J.K. TSCHANGO, Introduction: Special issue on geo-spatial information sharing and standardization *Zorica Nedovic-Budic* and *Tschangho John Kim*, *The Annals of Regional Science*, 1999, vol. 33, issue 2, pages 141-143.

[NED 00] Z. NEDOVIC-BUDIC, J.K. PINTO, "Information sharing in an interorganizational GIS environment", *Environment and Planning B: Planning and Design*, vol. 27, 455-474, 2000.

[NED 01] Z. NEDOVIC-BUDIC, J.K. PINTO, "Organizational (soft) GIS interoperability: Lessons from the U.S.", *International Journal of Applied Earth Observation and Geoinformation*, vol. 3, 290-298, 2001.

[OBE 95] N.J. OBERMEYER, "Reducing inter-organizational conflict to facilitate sharing geographic information", in H.J. ONSRUD, G. RUSHTON (eds.), *Sharing Geographic Information*, New Brunswick, NJ, Center for Urban Policy Research, 1995, 138-148.

[ONS 95] H.J. ONSRUD, G. RUSHTON (eds.), *Sharing Geographic Information,* New Brunswick, NJ, Center for Urban Policy Research, 1995.

[ONS 01] H.J. Onsrud, *Survey of National and Regional SDI Activities Around the Globe*, available at: http://www.spatial.maine.edu/~onsrud/GSDI.htm, 2001.

[OTO 84] L.J. O'TOOLE Jr., R.S. MONTJOY, "Interorganizational policy implementation: A theoretical perspective", *Public Administration Review*, 491-505, 44(6), 1984.

[PIN np] J.K. PINTO, Z. NEDOVIC-BUDIC, L. WARNECKE, "GIS database development and exchange: Interaction mechanisms and motivations", *Journal of the Urban and Regional Information Systems Association,* Forthcoming.

[PIN 95] J.K. PINTO, H.J. ONSRUD, "Sharing geographic information across organizational boundaries: A research framework", in H.J. ONSRUD, G. RUSHTON (eds.), *Sharing Geographic Information*, New Brunswick, NJ, Center for Urban Policy Research, 1995, 44-64.

[POR 98a] H. PORNON, *SIG, Pouvoir et Organisations: Géomatique et Stratégies D'acteurs*, Paris, L'Harmattan, 1998.

[POR 98b] H. PORNON, *SIG, et Organisations*, study report published by CERTU (Ministère de l'Equipement), Lyon, 1998.

[POR 98c] H. PORNON, *Partenariats Relatifs aux SIG*, study report published by CERTU (Ministère de l'Equipement), Lyon, 1998.

[POR 02a] H. PORNON, X. DUPASQUIER, *Les SIG Dans les Petites Communes*, Géomatique Expert, 2002.

[POR 02b] H. PORNON, "Inter-organizational partnerships and geographical information", *Proceedings from URISA Conference*, Chicago, 2002.

[ROC 99] S. ROCHE, J.-B. HUMEAU, "GIS development and planning collaboration: A few examples from France", *Journal of the Urban and Regional Information Systems Association (URISA)*, vol. 11, 5-14, 1999.

[ROC 03] S. ROCHE, C. CARON, "Local data policies and NSDI: Complementarity or antagonism?", *Proceedings from AGILE 2003*, Lyon, 2003.

[SOM 99] R. SOMERS, "Framework data survey – Preliminary report", *GeoInfo Systems*, vol. 9, 9(9), 1-35, 1999.

[STE 71] L.W. STERN, C.S. CRAIG, "Interorganizational data systems: The computer and distribution", *Journal of Retailing*, vol. 47, 73-91, 1971.

[TUL 01] D. TULLOCH, J. FULD, "Exploring county-level production of framework data: Analysis of the national framework data survey," *Journal of the Urban and Regional Information Systems Association*, vol. 13, 11-21, 2001.

[WAR 98] L. WARNECKE, J. BEATTIE, K. CHERYL, W. LYDAY, S. FRENCH, *Geographic Information Technology in Cities and Counties: A Nationwide Assessment*, Washington, DC, American Forests, 1998.

[WAR 99] L. WARNECKE, "Geographic information technology institutionalization in the nation's states and localities", *Photogrammetric Engineering and Remote Sensing*, vol. 65, 1257-1268, 1999.

[WES 91] A. WESTIN, Presentation, *Electronic Democracy Conference*, 1991.

Conclusion

This book was our opportunity to highlight the variety of organizational, social, and, in some ways, technological challenges linked to the development of the geographic information society. It also made it possible to understand in a more precise way the complexity of its operation. The reflections developed throughout the chapters have made it possible to clarify the strong interactions occurring between human, organizational, and social geographic information science (GIS) dimensions. It is the structures of the geographic information society in its whole that can be more globally interpreted.

The reader has probably noticed, while reading the book, that even though the chapters propose relevant paths for solutions and explanations, the reflections developed by the different authors still raise many questions concerning these interactions, despite the volume and diversity of the content. This observation motivated the development of a complementary reflection to formulate relevant paths to follow to continue research and development in this field. The contribution of several international experts to this book was precious, not only to report on the relations between GIS, organizations, and society but also to highlight common key research perspectives.

Written by Stéphane ROCHE, Claude CARON and Carine PÉRIBOIS.

The following question was asked of the different authors of this book during the SOS-IG workshop in Angers (France) from May 22nd to the 24th in 2003: *In your opinion, what research and development themes should be a priority during the next few years on the theme "Society, organizations and GIS?"* The results were particularly interesting. They represented more than fifty themes deemed relevant by all. Once categorized in decreasing order of importance, these themes were the subject of a selection based on consensus. The priority research themes were thus identified for development in the next few years. In conclusion, we give a global presentation of these themes updated for this second English edition.

One of the research priorities identified consists of analyzing the methods (and capabilities) of accessing geographic information and associated geographic information technologies (GIT). The question of the level of awareness and training offered to potential users was also discussed. We are also striving to better understand the socio-spatial effects of the diversification and democratization of GIS use.

Another problem was based on the necessity of understanding the role of geographic information in new forms of government (*e-government*) and making participatory decisions in the territory. The objective consists of evaluating decisional-process reengineering and modes of governing in light of GIT development. In terms of research and development, the necessity of developing collective geographic services to support the negotiation processes (process sharing) also was identified as a priority.

The improvement or the development of new forms of spatial representations produced and carried out by GITs to facilitate the process of appropriation of territorial problems also appeared as an topic of relevant reflection. This logically leads to a recognition of the importance of better understanding how the business visions linked to GIS are integrated, as well as a better understanding of the large variety of cognitive territory models.

In another venue, an identification and detailed analysis of the different variables conditioning the adoption of GITs appeared as one of the directions of priority research: social variables, cultural variables, spatial variables, etc. More globally, the idea was to attempt to understand how the new social practices inherent to the geographic information society are structured by considering different parameters: ethnic groups and types, religions, etc.

Another axis revealed to be important consisted of the determination of the level of use, as well as of organizational use, of GITs (goals set, existing process integration, costs and benefits, etc.). In particular, in the current context, we need to understand the capacity of GISs to generate strategic and competitive advantages in a business perspective.

The roles (official or unofficial) played by GITs, as well as the value (qualitative or quantitative) of these technologies must also be evaluated in a more precise way in the contexts of participatory methodologies mobilizing the stakeholders of territorial development.

The history of technology in general, and of GIT in particular, is also a scientific challenge. In particular, research involving the processes of social construction of these technologies should be supported and put into perspective in the different contexts and social stakes. This type of reflection would assuredly help us in taking a long enough step backward to highlight the global evolution of GIT and thus understand the future of the geographic information society.

The theme of public participation that mobilizes GITs also came out as a priority. This theme is vital in improving our understanding of the methods and technologies we will adapt to the implementation of participatory GIS (PPGIS or public participation GIS) for its efficient evolution. Beyond the development of generic methods and techniques, these reflections highlighted the importance of developing a theoretical “contextualized” scope to develop PPGIS.

The study of the organizational and social impact of technological ubiquity is also considered a major priority. The effects of GIT omnipresence are important challenges for the development of the geographic information society. Understanding the human and social impacts of mobile and wireless technologies, for example, is a major objective of research to develop, the theme of "GIS and society." More globally, the effects of these technologies on the evolution of the GIS sector, as well as the impact of the geographic information society as a whole, must be analyzed to better understand these effects and be able to anticipate the coming changes.

Another thematic priority identified by the authors of this book centered on the improvement of current implementation methods for information systems (particularly geographic) for their efficient integration with human, social, and (contextual) geographic dimensions of development projects. We must better understand the non-technical aspects of the development and implementation of GIT.

Web 2.0 development and its geospatial component, GeoWeb, centered on the concept of collective intelligence and characterized by a greater openness and the idea of sharing and participation, reiterates a series of transversal reflections determined by our research question. One of the current research priorities identified was measuring the effects of Geoweb on the development of the geographic information society. How does it renew the problem of social interactions between GIS and societies?

The democratization process that comes with GeoWeb is undoubtedly one of the most important elements identified. Google Earth and Map (and other virtual maps) are the most obvious materializations of this process. More globally, this process is characterized by new practices (volunteered geographic information [VGI], geotagging, geocrowdsourcing, mass public GPS navigation, geocaching, etc.); new technologies (geoblog, wikiGIS, Application Programming Interface, Keyhole Markup Language, etc.); and new information (points of interest, etc.). The questions resulting from these lines of reflection are numerous. But those questions involving quality (new metrics) and the socio-economic value of these new geographic contents generated consciously by users are topical. Scientific questioning raised in the context of citizen-sensors and the concept of citizen science, as evoked by M. Goodchild, are also relevant research paths in the attempt to better understand the complex process of development in the geographic information society.

These research themes provide a clear and particularly synthetic range of priority questions involving the geographic information community, especially the community interested in organizational GIS aspects. These themes are not mutually exclusive. There are partial intersections and cross-checks among them. From this vantage, however, it becomes even easier to grasp the scope of the questioning that remains to be addressed. It is vital to pursue research actions involving multi-disciplinary and multi-cultural expertise to complete the overview presented in this book on the geographic information society. To do this, we must maintain the view that supporting exchanges and the sharing of knowledge and know-how between the human and social sciences and among information and communication techniques and sciences is very important. Only by favoring such rich and eclectic mixes of knowledge can we better understand the organizational and social challenges of GISs as a whole.

List of Authors

François BERGERON
Département des Systèmes d'Information Organisationnels
Faculté des Sciences de l'Administration
Laval University
Quebec
Canada

Claude CARON
Département SIMQG
Groupe de recherche en Géobusiness
University of Sherbrooke
Quebec
Canada

Jean-Jacques CHEVALLIER
Département des Sciences Géomatiques
Faculté de Foresterie et de Géomatique
Laval University
Quebec
Canada

Nicholas CHRISMAN
Department of Geography
State University of Washington
Seattle
USA

Grégoire FEYT
Institut de Géographie Alpine
Joseph Fourier University
Grenoble
France

Rina GHOSE
Department of Geography
University of Wisconsin
Milwaukee
USA

Lin GINGRAS
Département SIMQG
Groupe de recherche en Géobusiness
University of Sherbrooke
Quebec
Canada

François GOLAY
Laboratoire de SIG (LaSIG)
École Polytechnique fédérale de Lausanne
Switzerland

Wladimir MAJOR
Division des Systèmes d'Information
State of Geneva
Switzerland

Zorica NEDOVIC-BUDIC
Department of Urban and Regional Planning
University of Illinois
Urbana-Champaign
USA

Carine PERIBOIS
Department of Geography
CARTA CNRS
University of Angers
France

Jeffrey K PINTO
Black School of Business
Penn State University
Erie
USA

Henri PORNON
IETI Consultants
Macon
France

Benoît RAVELEAU
CERIPSA
Université Catholique de l'Ouest
Angers
France

Stéphane ROCHE
Département des Sciences Géomatiques
Centre de Recherche en Géomatique
Faculté de Foresterie et de Géomatique
Laval University
Quebec
Canada

Robin SMITH
Department of Town and Regional Planning
University of Sheffield
UK

Index

W